PRACTAMATICS AND ARITHMAGIC

by

Anthony P. Mazalas

A Hearthstone Book

Carlton Press, Inc.　　　　　New York, N.Y.

CONTENTS

MULTIPLICATION

Method

DIVISION

POWERS OF TEN

PREFACE

Throughout the centuries, man has devised, invented, and per-
fected ways and methods to perform the tasks of arithmetic such
as multiplication, division, addition, and subtraction. Included
among these were finger counting, using pebbles and sticks, the
abacus, Roman numerals, pencil and paper, the slide rule, me-
chanical calculators and finally the electronic calculator. There
are many more methods not listed above. Some were tedious,
some time consuming, some difficult to understand and perform,
some rapid and some called short-cuts or speedy math.

The author does not intend this volume to be a history of the
art of calculation, but has chosen one aspect to examine, calculation with paper and pencil, and mentally. The author hopes the
reader will enjoy reading this first volume and try the various
ways outlined. It is also the wish of the author that this book may
help the reader to perfect other ways of multiplying and dividing
so that he or she may become more proficient.

Some of these methods appear for the first time in print. A
bibliography appears at the end of the volume for further ref-
erence.

It is also the intent of the author to compile functions of arithmetic and mathematics other than those covered in this volume.

May the reader find enjoyment and pleasure and further
achievement with this book.

INTRODUCTION

The only mathematical ability necessary to understand and use this volume is the capabilities of basic addition, subtraction, multiplication and division. We learned the multiplication table to twelve when we were young. In fact, it was drilled into us so we memorized the table. It may have been better if we learned the table to a higher figure instead of twelve times twelve. Actually, the need for higher multiplication tables is not necessary. It will be demonstrated later how easy it is to multiply 25 × 46, etc.

There are no gimmicks presented here and all methods are based on mathematical proof and procedures. Some of these methods may be considered as short-cuts. The author does not include the proofs for the various methods. The proofs can be verified from the bibliography at the end of the volume.

The methods presented do not follow any definite order or pattern. Some are simpler than others. Some may appeal to certain people, some to other people. Many of these methods can be performed mentally without using pencil and paper.

A quick introduction to multiplication:

Standard Method

```
       47          1.)  2 x 7 = 14, set down 4, carry 1
     x 52          2.)  2 x 4 =  8 + 1 carry = 9
       94          3.)  5 x 7 = 35, set the 5 down (one place
   +  235                        to left), carry 3
     2444  Ans.    4.)  5 x  4 = 20 + 3 carry + 1 from
                                 the 9 + 5 addition.
```

There are many chances to make errors, especially if the multiplication is more complex.

A simple method called cross-multiplication:

```
            Steps
     47      1.)  2 x 7 = 14, set down
   x  52          5 x 4 = 20, set down
   2014      2.)  (5 x 7) + (2 x 4) = 35 + 8 = 43
 +   43           Add mentally
   2444   Ans.
```

A quick glance through the book will show methods where 4-digit numbers are multiplied with accuracy and speed.

The cubing of numbers is included but it is kept to a minimum since the average need for cubing is rather rare.

Comments on Division

In a great many instances, division can be as simple as multiplication. No other mathematical operation is disliked by so many as division. But division can be accomplished by factoring, reduction, subtraction, boundaries, aliquot parts, etc. There is a summary of divisional rules (if one number can be divided by another number). Also included are methods for checking division results.

A chapter is included on powers of 10 or engineering math. The use of powers of 10 is very handy in calculating with very large or very small figures. Using some of the methods described it is possible to multiply millions by billions mentally with an accuracy of 1%.

The various examples are self-explanatory, whereas the rules alone may be difficult to interpret for some. Notice that there are no practice problems included. The reader can create problems on his or her own after following the examples.

In calculation it is very important to examine the numbers carefully instead of plunging into the task of calculation immediately. You will be able to recognize the relationship that exists between numbers.

Note also that many of the calculations are performed from left to right instead of from the traditional right to left procedure. Some of the examples seem drawn out and lengthy. That is, because each step is shown carefully and not left to the reader to figure out where the numbers came from. No doubt the reader will use a shorter form in calculating and will not write down each step.

The reason for the "arithmagic" in the title is that to many, some of the methods appear as if by magic. There is no magic in calculation. There have been and will be math prodigies. When asked how they can calculate so quickly, they cannot explain their procedures.

By no means does this volume include all the methods used. Yet to many, some of the ways will be new, as this is the first time they appear in print.

This book is recommended to math teachers who would like to break up the monotony of teaching math. If one knew every method perfectly, he or she would be a genius.

The author hopes that this volume will bring pleasure to those who use it and will also help them in achieving higher skills and goals.

PRACTAMATICS AND ARITHMAGIC

MULTIPLICATION

1. Cross Multiplication

Rule: Multiply the units of each number. If the product is one digit, add a zero in front. Then multiply the tens (and hundreds) of each number. Set down the answers. Multiply the units of one number by the tens (and hundreds) of the other number. Repeat using the units of the second number by the tens (and hundreds) of the first number. Add the results to the center of the first multiplication.

1.) a. 3 6 b.
     c. 7 4 d.
         2124
          42
       + 12
         2664

1.) a x c   3 x 7 = 21
2.) b x d   6 x 4 = 24
3.) c x b   7 x 6 = 42
4.) d x a   4 x 3 = 12
5.) set down 1.) and 2.). add steps
     3.) and 4.) to the center of 1.) and 2.)

2.)   6 2
     x 3 4
       1 8 0 8
     +   3 4
       2 1 0 8

2 x 4 = 08
3 x 6 = 18
4 x 6 = 24
3 x 2 = 06

3.) 1 2 3
     x   3 4
       3 6 1 2
         4 8
     +   0 9
       4 1 8 2

3 x 12 = 36
3 x 4 = 12
4 x 12 = 48
3 x 3 = 09

4.)   1 2. 6
     x 1 0. 5
       1 2 0 3 0
           6 0
     +   6 0
       1 3 2.3 0

12 x 10 = 120
5 x 6 = 30
10 x 6 = 60
5 x 12 = 60
point off two places

5.)   1 2 6
       x 1 0 5
       1 2 0 3 0
             6 0
             6 0
         1 3 2 3 0

10 x 12 = 102
5 x 6 = 30
6 x 10 = 60
5 x 12 = 60

17

2. Average Mean, Part 1
(when difference between numbers is 1)

Rule: When the difference between two numbers is one, square the average mean and subtract one from the answer.

1.) $29 \atop x \ \underline{31}$ } 30 (average mean) 30^2 = $\begin{array}{r} 900 \\ - \quad 1 \\ \hline \end{array}$

answer   899

2.) $39 \atop x \ \underline{37}$ } 38 38^2 = $\begin{array}{r} 1444 \\ - \quad 1 \\ \hline \end{array}$

answer   1443

3.) $101 \atop x \ \underline{99}$ } 100 100^2 = $\begin{array}{r} 10000 \\ - \quad 1 \\ \hline \end{array}$

answer   9999

4.) $303 \atop x \ \underline{305}$ } 304 304^2 = $\begin{array}{r} 92416 \\ - \quad 1 \\ \hline \end{array}$

answer   92415

5.) $5.4 \atop .56$ } 55 55^2 = $\begin{array}{r} 3025 \\ - \quad 1 \\ \hline \end{array}$

Point off 3 places   answer   3.024

6.) $21.5 \atop x \ \underline{.216}$ 214 214^2 = $\begin{array}{r} 45\,796 \\ - \quad 1 \\ \hline \end{array}$

Point off 4 places   answer   4.5795

This method becomes difficult when numbers with three digits are multiplied. It is harder to square a three-digit number. An alternate method is shown in Method 3.

3. Average Mean, Part 1
(alternate Method)

Rule: Square the hundreds digit (leave out any tens or units digits). Square the means. Set down under the square of the hundreds digit. Double the means digit and multiply by the single hundreds digit. Set down. Add these three numbers and subtract one from the sum.

1.)
$$\left.\begin{array}{r} 475 \\ x\ 477 \end{array}\right\}\ 76\ \text{(mean)}$$

Square hundreds	400^2 =	160000
Square means	76^2 =	5776
Double means. Multiply hundreds. (76)(2)(400)	= +	60800
		226576
	−	1
Answer		226575

2.)
$$\left.\begin{array}{r} 103 \\ x\ 105 \end{array}\right\}\ 04$$

100^2	=	10000
04^2	=	16
(04 x 2)(100)	= +	800
		10816
	−	1
Answer		10815

3.)
$$\left.\begin{array}{r} 1029 \\ x\ 1031 \end{array}\right\}\ 30$$

1000^2	=	1000000
30^2	=	900
(30 x 2)(1000)	=+	60000
		1060900
	−	1
Answer		1060899

4.)
$$\left.\begin{array}{r} .519 \\ x\ .521 \end{array}\right\}\ 20$$

500^2	=	250000
20^2	=	400
(20 x 2)(500)	= +	20000
		270400
		1
answer	−	
Point off 6 places.		270399

4. Average Mean, Part 2
(when the average mean is equi-distant by more than 1)

Rule: Square the average mean. Square the equi-distant figure and subtract from the mean squared for the answer.

1.)
$$\left.\begin{array}{r} 77 \\ x\ 73 \end{array}\right\}\ \begin{array}{l} 75\ \text{(average mean)} \\ \pm 2\ \text{(difference)} \end{array}$$

square 75	=	5625
square 2	= 1−	4
answer		5621

2.) 104 ⎫   100 100^2 = 10000
  x 96 ⎭ ± 4 4^2 = -   16
                answer   9984

3.) 112 ⎫ 100 100^2 = 10000
  x 88 ⎭ +12 12^2 = - 144
          answer   9856

4.) 11.2   Disregard the
   x .88   decimals. Insert
    9.856   3 places in answer

5.)   37   30 30^2 = 900
   x 23   7 7^2 =- 49
          answer   851

When the mean is not equi-distant from both numbers, this method still applies. However the mean to be squared is not a whole number, but a decimal, making it more difficult to square. An alternative procedure is shown in Method 5.

6.)   78   75.5 (average mean) 75.5^2 = 5700.25
   x 73   2.5 (difference) 2.5^2 = - 6.25
                             answer   5694

5. Average Mean, Part 2
(When average Mean Is Not Equi-distant From Both Numbers)

Alternate Method

Rule: Divide the difference between the two numbers by 2. Drop the decimal. Add the whole number to the lower multiplier. Square. Square the difference (without decimal) and subtract from above. Add the lower number to the result for the answer.

1.) 76 ⎫ 7 (difference)   7 ÷ 2 = 3.5 (drop the .5)
   x 69 ⎭   add the 3 to 69; 69 + 3 = 72
                            Square 72   = 5184
              Square 3 and subtract = - 9
                                       5184
                            Add 69   + 69
                                       5244   Ans.

2.) $\begin{array}{r} 84 \\ \text{x } 51 \end{array}\Big\}$ 33 (difference)

1/2 of 33 = 16.5 Drop the .5
51 + 16 = 67

$$67^2 = 4489$$
$$-(16)^2 = \underline{-256}$$
$$4233$$
$$+\ 51 \quad + \underline{\quad 51}$$
$$4284 \quad \text{Ans.}$$

3.) $\begin{array}{r} 214 \\ \text{x } 211 \end{array}\Big\}$ 3 (difference)

1/2 of 3 = 1.5. Drop the .5
211 + 1 = 212

$$212^2 = 44944$$
$$-(1)^2 = -\underline{\quad 1}$$
$$44943$$
$$+\ 211 = +\underline{\ 211}$$
$$45154 \quad \text{Ans.}$$

6. Rounding Off to Easier Numbers

Rule: Round off one of the numbers to an easier number to multiply. Multiply this number by the other number. Add or substract the amount lowered or raised the number you rounded off.

1.)
$$\begin{array}{r} 99 \\ \text{x } 67 \end{array} \qquad \begin{array}{r} 100 \\ \text{x } 67 \\ \hline 6700 \\ -\ \underline{67} \end{array}$$
$$\text{Answer} \quad 6633$$

Round off 99 to 100. Multiply by 67. Since you multiplied 67 one more time, (by 100 instead of 99), you must subtract a 67 from 6700.

2.)
$$\begin{array}{r} 47 \\ \text{x } 21 \end{array} \qquad \begin{array}{r} 47 \\ \text{x } 20 \\ \hline 940 \\ +\ \underline{47} \end{array}$$
$$\text{Answer} \quad 987$$

Round off 21 to 20. Since you multiplied 47 one less time by 20 instead of 21, you must add 47 to the above sum.

3.)
$$\begin{array}{r} 124 \\ \text{x } 98 \end{array} \qquad \begin{array}{r} 124 \\ \text{x } 100 \\ \hline 12400 \\ -(124\ \text{x2}) -\ \underline{248} \end{array}$$
$$\text{Answer} \quad 12152$$

4.)
$$\begin{array}{r} 456 \\ \text{x } 103 \end{array} \qquad \begin{array}{r} 456 \\ \text{x } 100 \\ \hline 45600 \\ +(56\text{x } 3) + \underline{1368} \end{array}$$
$$46968$$

6.) $\begin{array}{r} 99.8 \\ \text{x } 3.24 \end{array}$

Disregard decimals. Perform as in example 5. Point off 3 decimal places to left in answer. 323.352

5.)
$$\begin{array}{r} 998 \\ \text{x } 324 \end{array} \qquad \begin{array}{r} 1000 \\ \text{x } 324 \\ \hline 324000 \\ -(324\ \text{x } 2) -\ \underline{628} \end{array}$$
$$\text{Answer} \quad 323352$$

21

7. Binary Method

Rule: Divide the multiplicand and double the multiplier until the sum of 1 is reached by the multiplicand. Disregard decimals or left overs. Cross out the even multiplicands (and their multipliers). Add the resulting multipliers to obtain the final answer. Note that it is easier to divide by 2 or to double.

1.)

	Multiplicand		Multiplier
Even	~~232~~	~~x~~	~~456~~
"	~~446~~	~~x~~	~~342~~
"	~~58~~	~~x~~	~~624~~
Odd	29	x	1248
Even	~~14~~	~~x~~	~~2496~~
Odd	7	x	4972
"	3	x	9984
"	1	x	19968
Answer			36192

2.)

43	x	17
21	x	34
~~10~~	~~x~~	~~68~~
5	x	136
~~2~~	~~x~~	~~272~~
1	x	544
Answe		731

3.)

.087	x	.36
87	x	36
43	x	72
21	x	144
~~10~~	~~x~~	~~288~~
5	x	576
~~2~~	~~x~~	~~1152~~
1	x	2304
		3131

4.)

~~42~~	~~x~~	~~17~~
21	x	34
~~10~~	~~x~~	~~68~~
5	x	136
~~2~~	~~x~~	~~272~~
1	x	544
Answer		714

Disregard the decimal point. Point off five places in above answer. (3 + 2 places)
Answer .03132

This illustrates the difference between multiplying 42 × 17 and 43 × 17.

8. Napier's Rods

Rule: Arrange a format of slanted lines as shown in the example (app. 60 degree slant). Multiply the multiplicand by each digit of the multiplier in turn. Enter the results as shown. Then add the slanted columns as shown for the answer.

Example: 76 × 38

```
        /5/4/          8 x 6 = 48    8 x 7 = 56
     2/1/6/8/          3 x 6 = 18    3 x 7 = 21
  + /1/8/
  ─────────               answer  2888
    2 8 8 8
```

1.)   326   3 x 6 = 18   3 x 2 = 06
    x 413   3 x 3 = 09
    134638   1 x 6 = 06   1 x 2 = 02
```
      //0/0/1/                 1 x 3 = 03
      //9/6/8/         4 x 6 = 24    4 x 2 = 08
    /0/0/0/                 4 x 3 = 12
    /3/2/6/
  /1/0/2/
 +/2/8/4/
 ──────────
  1 3 4 6 3 8  Ans.
```

Note: When a product has a single digit add a 0 in front.

2.) .095
   x .0079
   .0007505
```
        /8/4/          9 x 5 = 45
     6/3/1/5/          9 x 9 = 81
   + /3/5/             7 x 5 = 35
   ─────────           7 x 9 = 63
     7 5 0 5
```

Count off 7 places to left, adding zeros where necessary.

9. Step-by-Step Breakdown

Rule: Take one of the numbers and break it down to hundreds, tens and units with single significant digits. Use these to multiply the other number and add the products.

1.) 326 x 413   2.) 56 x 42

     326 x 400 = 130400   56 x 40 = 2240
     326 x 10 =   3260   56 x 2 = 112
     326 x   3 =   978   2352   Ans.
                 134638 Ans.

3.) 371 x 392   4.) .124 x .38
     371 x 300 = 111300   124 x 30= 3720
     371 x 90 = 33390   124 x 8 = 992
     371 x   2 =   742   4712   Ans.
                 145432

23

5.) 4250 x 3986   Change to 3986 × 4250
     3986 x 4000 = 15944000
     3986 x 200 =   797200
     3986 x   50 =   199300
     3986 x   0 =   0
                  16940500 Ans.

6.) 129 x .567
     129 x 500 = 64500   Point off 3 places
     129 x 60 = 7740   Answer   73.143
     129 x   7 =   903
                 73143

10. Algebraic

Rule: Multiply the tens of each number. Then multiply the tens of the first number by the units of the second number. Set down. Multiply the units of the first number by the tens of the second number and add to first answer. Multiply the units and add to above, one place to the right.

Example
          a 33 b   (c x a) 2 x 3 = 6
       X c 24 d   (c x b) 2 x 3 = 6
             66   (d x a) 4 x 3 = 12
             12   (b x d) 4 x 3 = 12
        +   12
            792 Answer

1.)   84
      x 27
         168   (2 x 8),(2 x 4)
          56   (7 x 8)
      +   28   (7 x 4)
        2268   Answer

2.)   23
      x 45
           8   (4 x 2)
     •   12   (4 x 3)
          10   (5 x 2)
      +   15   (5 x 3)
        1035   Answer

3.)   37
      x 56
          15
          35
          18
      +   42
        2072   Answer

4.)   324
      x 215
         648
         324
          15
          10
      +   20
       69660   Answer

*Note how the 12 is placed. (not next to the 8). The tens digit must be under the units digit of the first multiplication.

24

11. By Factoring

Rule: If one of the numbers can be factored, multiply the other number by the factors in turn.

Example:

```
        17        12 is (4 x 3)      17 x 4 =  68
      x 12                           68 x 3 = 204   Answer.
```

```
1.)   32          24 = (6 x 4)
    x  24          32 x 6 =    192
                  192 x 4 =    768  Ans.
```

```
2.)  326    88 = (11 x 8)            3.)   978    48 = (6 x 8)
   x  88    11 = (10 + 1)                x  48    978 x 6 =  5868
            326 x  10  =  3260                    5868 x 8 = 46944  Ans.
          + 326 x   1  = +  326
                          3586
        3586 x 8  =  28688   Ans.
```

```
4.)  .098 x .036                 5.)   756    216 = (6 x 6 x 6)
  or   98 x 36                       x 216    756 x 6 =  4536
       36 = 6 x 6                             x6    =x     6
       98 x 6 =    588                              27216
      588 x 6 =  3528                         x6     x    6
      point off 6 places                           163296  Ans.
       .003528  Ans.
```

12. Double and Half

Rule: Double one number and halve the other. Repeat until you end up with one digit for one of the numbers. Multiply the other number by this digit for answer.

```
1.)      326 x 128          2.)    714 x 34
         652 x  64               1428 x 17
        1304 x  32               2856 x 8.5
        2608 x  16             x    8
        5216 x   8             22848     plus 1/2 of 2856
      x      8               +  1428
        41728   Ans.           24276   Ans.
```

25

3.)   134 x 2.5   4.)   176 x 2.25 (21/4)

```
3.)      134 x 2.5          4.)    176 x 2.25 (21/4)
          67 x 5                    88 x 4.5
      x    5                        44 x 9
         335   Ans.               x   9
                                   396   Ans.


5.)     .33 x .68           6.)     413  x   26
    or   33 x  68                   826  x   13
         66 x  34                  1652  x   6.5
        132 x  17                 x    6
        264 x   8.5               9912     plus 1/2 of
    x     8                     +  826     1652
       2112    plus 1/2 of      10738  Ans.
    +   132     264
       2244
count off 4 decimal places
    .2244 Ans.
```

13. Medium and Large Multiplication Tables
(Multiplier is a single Digit)

Rule: Multiply the tens. Multiply the units. Add (mentally).

```
1.)   17 x 8;   (8 x 10) + (7 x 8) = 80 + 56 = 136   Ans.
2.)   19 x 9;   (9 x 10) + (9 x 9) = 90 + 81 = 171   Ans.
         or say 90 + 81 = 171
```

Large Multiplication Table
(11 × 11 to 19 × 19)

Rule: Add the units of one number to the other number. Multiply sum by 10. Add on the units product.

```
3.)  13 x 18               4.)   19 x 19
     13 + 8  = 21                19 + 9 = 28
        21 x 10 =  210           28 x 10 =  280
     + (3 x 8) = + 24             9 x  9 = + 81
                 234                        361   Ans.
```

For 21 x 21 to 29 x 29

```
5.)  22 x 21               6.)   27 x 29
     22 + 1 = 23                 27 + 9 = 36
        23 x 20 = 460            36 x 20 =  720
     +(2 x 1) = + 2             +(7 x 9) = + 63
                462   Ans.                  783   Ans.
```

26

The same method is used for 31 × 39 to 91 × 99. Multiply by the tens digit 30 to 90.

14. Multiplications By Aliquot Parts

25 is an aliquot part of 100 or ¼ of 100
12.5 is an aliquot part of 100 or ⅛ of 100
 6.25 is ¹⁄₁₆ of 100

Rule: Divide the number of aliquot parts that the multiplier is and multiply that number.

1.)   384 x 25;   384 ÷ 4 = 96;   96 x 100 = 9600   Ans.

2.)   84 x 12.5;   84 x 1/8 = 10.5;   10.5 x 100 = 1050   Ans.

3.)   64 x 6.25;   64 x 1/16 or $\frac{64}{16}$ = 4   4 x 100 = 400   Ans.

Multiplying by numbers near an aliquot number such as 11½, 13½, etc.

4.)   440 x 11 $\frac{1}{2}$ 440 x 1/8 x 100   =   5500
      below   subtract   -   440
                                               5060   Ans.

5.)   440 x 13 $\frac{1}{2}$ $\frac{440}{8}$ = 55 x 100 =   5500
      above   add   +   440
                                               5940   Ans.

6.)   440 x 14 $\frac{1}{2}$ $\frac{440}{8}$ = 55 x 100 =   5500
      above   add 2 x 440   +   880
                                               6380   Ans.

Note: 12 $\frac{1}{2}$ is 1/8 of 100

15. Simplifying the Multiplier

When a multiplier is of such a nature that part of it may be taken as an exact multiple of another digit in the multiplier, such as 186, which can be written as 6 × 3, 6; or 856, which can be written as 8, 8 × 7; etc.

When the factors are the first two digits:

Rule: Multiply the number by the common factor. Set down. Then multiply this semi-product by the odd factor. Set down underneath the first product, moving it one place to the left. Then add for the answer.

```
1.)    2574          186 is 6 x 3, 3
     x   186          2574 x 6  =  15444
                      15444 x 3 + 46332
                                  478764   Ans.

2.)     837           273 is 9 x 3, 3
     x   273           837 x 3  =   2511
                       2511 x 9  = 22599
                                  228501   Ans.
```

When the factors are the last two digits:
Move the second semi-product one place to the right and add.

```
3.)   5462           856 is 8, 8 x 7
    x  856           5462 x 8  =    43696
                     43696 x 7 = +  305872
                                  4675472   Ans.

4.)   7943           728 is 7, 7 x 4
    x  728           7943 x 7  =   55601
                     55601 x 4 = +  222404
                                 5782504   Ans.
```

When one of the numbers has two identical factors:

```
5.)    1456          36 is 6 x6   twice
     x 3636          1456 x 6  = 8736
                     8736 x 6  = 52416
```

Add 52416 to itself but move it over two places to the right.

```
           52416
     +       52416
           5294016   Ans.

6.)   6789          24 is 8 x 3 twice
    x 2424           6789 x 8 = 54312
                     54312 x 3 = 162936
```

Add this to itself but move it over two places to the right.

```
           162936
     +       162936
           16456536   Ans.
```

28

7.) This method can simplify problems like this one.

$$\begin{array}{r} 24246 \\ \times\ \underline{\quad 6789} \end{array}$$

Do this example as above.
To the answer above add the product of 6789 × 6 two places to the right.

```
16456536   Answer from above
+    40734   6789 x 6 = 40737
164606094   Ans.
```

16. Sum by Difference

When numbers to be multiplied can be expressed as the sum and difference between two numbers, the product equals the square of the first minus the square of the second. 73 × 67 may be expressed as 70 + 3 multiplied by 70 − 3. The product equals 70 × 70 − 3 × 3 or 4900 − 9 = 4891.

Rule: Express both numbers as a simple number easily squared. One number will be so many digits above and the other number will be the same number of digits below. Square the sum and difference and subtract from the common square.

```
1.)   132 x 128   (130 + 2) x (130 - 2)
                  (130)² - (2)²
                  16900 - 4 = 16896   Ans.
2.)   1412 x 1388  (1400 + 12)(1400 - 12)
                  (1400)² - (12)²
                  1960000 - 144 = 1959856   Ans.
```

When the sum and difference are not equal, the positive difference is greater than the negative difference.

Rule: Square the boundary. Multiply the boundary by the positive and by the negative differences. Subtract the product of the differences and add to the difference of boundaries. Then, since this is a positive difference, you add this to the squared boundary. When the negative difference is greater, add the negative difference to the negative product to obtain a larger negative answer. Subtract this (add a negative number) from the squared boundary.

29

3.)   63 x 58   4.)   72 x 67

```
3.)      63 x 58                    4.)     72 x 67
     60 + 3        3 x 60 = 180     70 + 2      70 + 2  =  140
   x 60 - 2     -( 2 x 60) =-120   x 70 - 3     70 - 3  = -210
   3600                       60   4900        4900        -70
                 2 x -3  = - 6                 2 x -3  =   - 6
                            54                              -76
     3600 + 54  = 3654   Ans.           4900 - 76 =  4824   Ans.


5.)    405 x 396                 6.)   1404 x 1393
    400 + 5      5 x 400 =  2000       1400 + 4     1400 x 4 = 5600
  x 400 -  4   -(4 x 44) = -1600     x 1400 - 7     1400 x-7 =-9800
  160000                    400      1960000                 -4200
               5 x -4   =  - 20                  4 x -7   = - 28
                           380                              -4228
      160000 + 380 = 160380   Ans.   1960000 - 4228 = 1955772  Ans.
```

17. Any Two-Digit Number by Any Three-Digit Number

Rule: Multiply unit digits. Answer must be in two-digit form. Circle last digit. Multiply tens digit by units digit of multiplier and add to this the units digit times tens digit of multiplier. Add any carry. Circle the last digit. Multiply the hundreds digit by units digit of multiplier. Add to this the product of tens digit and the tens digit of multiplier. Circle last digit. Multiply hundreds digit by tens digit of multiplier and add any carry. Circle both digits. The answer is in the circles.

```
1.)
         724       2 x 4                          = 0 ⑧
       x  32       (2 x 2) + (3 x 4) + 0 (carry)  = 1 ⑥
                   (2 x 7) + (3 x 2) + 1 (carry)  = 2 ①
                   (3 x 7) + 2(carry)             = ②③
                                  23168  Ans.

2.)      856       6 x 9                          = 5 ④
       x  49       (9 x 5) + (4 x 6) + 5 (carry)  = 7 ④
                   (9 x 8) + (4 x 5) + 7 (carry)  = 9 ⑨
                   (4 x 8) + 9 (carry)            = ④①
                                  41944   Ans.
```

This method can be used with multiplicands of four, five or more digits.

3.) 1234
   x 56

6 x 4   = 2④
(6 x 3) + (5 x 4) +2 (carry)   = 4⓪
(6 x 2) + (5 x 3) +4 (carry)   = 3①
(6 x 1) + (5 x 2) +3 (carry)   = 1⑨
(5 x 1) + 1(carry)   = ⓪6

                          69104   Ans.

1234
56

1234
56

1234
56

18. Cross-Multiplication (Another Method)

Rule: Round off numbers to the tens. Multiply and record. Multiply the units. Add to above at right. Cross multiply digits and add as shown in examples. Answer is the sum of all the products.

1.)	64	= 60			2.)	91		90
	x 40	x 40				x 32		x 30
		2400						2700
	0 x 4 =	0				2 x 1 =		02
	4 x 4 =	16				3 x 1 =		03
	0 x 6 =	0				9 x 2 =	+	18
		2560	Ans.					2912 Ans.

3.)	143	140		4.)	12.6	120
	x 128	x 120			x 3.2	x 30
		16800				3600
	8 x 3 =	24			2 x 6 =	12
	12 x 3 =	36			3 x 6 =	18
	14 x 8 = +	112			2 x 12 = +	24
		1 8304 Ans.				4032

Count off two places
40.32   Ans.

19. Numbers Separated by Two Digits

Rule: Square the middle term. Subtract 1 from the squared product.

1.)   29 30^2 = 900   2.)   89   • 88^2 = 7744
   x 31   - 1   x 87   - 1
                  899   Ans.   7743   Ans.

                                   • 88^2
3.)   201 202^2 = 40804 90^2 = 8100
   x 203   - 1   - 8^2 x 4 - 356
               40803   Ans.   7744   Ans.

4.)   1307 1306^2 = 1705636 1306^2 130^2 = 16900
   x 1305   - 1 6^2 =   36
            1705635   Ans.   6 x 13 x 2 156
                                   1705636   place

Numbers Separated By Three Digits   156 in center

Rule: Square the higher number. Subtract three times this higher number from the square for the answer.

5.)   29 32^2 =   1024   6.)   128 131^2 = 17161
   x   32   -3 x 32= - 95   x 131   -3 x 131 = - 393
               928   Ans.   16768 Ans.
                            131^2 13^2 = 16901 1^2 = 01
6.)   756 759^2 = 576081   13 x 1 x 2 + 26
   x 759   -3x 759 = - 2277   17161
                573804   Ans.

   759^2
   75^2 = 562581 ← 9^2
75 x 9 x 2 + 1350
       576081

20. Numbers Whose Unit Digits Are Alike

Rule: Square the units. Set the unit product down and carry the tens unit, if necessary. Add the tens and multiply by the like unit. Add any carry. Multiply the tens digits and add any carry.

1.)   72 2^2 = ④
   x   42   (7 + 4)= 11 x 2 = 2②
        7 x 4 = 28 +2(carry) = ㉚
           3024   Ans.

2.)   413   13 x 13   = 1 (6 9)
   x 813   Set down 69 and carry 1
               8 + 4 = 12, 12 x 13 = 156 + 1 Carry
               Set down 57, carry 1
                                                           = 1 (5 7)
               8 x 4 = 32 + 1 carry =
                                                             (3 3)
                      3357 69   Ans.

   3.)   99   9 x 9 = 8 (1) Set down 1, carry 8
      x 89   9 + 8 = 72 x 9 = 153 + 8 carry = 1 6 (1)
                 Set down 1, carry 16
                 9 x 8 = 72 + 16 carry = (8 8)
                        8811   Ans.

   4.)   189   9 x 9 = 81. Set down 1 and carry 8   8 (1)
      x   39   18 + 3 = 21 x 9 = 189 + 8 =   19 (7)
                 18 x 3 = 54 + 19 = 73   (7 3)

                                          7371   Ans.

21. Numbers Whose Tens Digits Are Alike

Rule: Multiply the units. Set down the unit digit of product and
carry (if necessary) the tens. Add the units. Multiply by the
like ten digit and add any carry. Set down the result.

   1.)   97   7 x 3 = 2 (1) set down the 1 and carry 2.
      x 93   7 + 3 = 10 x 9 = 90 + 2 (carry) = 9 (2)
                 Set down the 2, carry the 9.
                 9 x 9 = 81 + 9 (carry) = (90)
                                          9021   Ans.

   2.)   763   First multiply 63 x 61
      x   261   1 x 3 = (3) Cross multiply 3 x 1 = 4 x 6 = 2 (4)
                   7 x 1 = 2 x 3 = 13   6 x 6 = 36 + 2 = (3 8)
        38 43   2 x 6 + 6 x 7 = 54   3843
        13   2 x 7 = 14
        54
     + 14
      199143   Ans.   cross multiply the hundreds
and units. Add. Cross multi-
ply the hundreds and units.
Add (one place to the left).
Multiply hundreds and add
product (one place to the left).

3.) 79 $9 \times 5 = 4$⑤
   x 75 $9 + 5 = 14 \times 7 = 98 + 4 = 1\,0$②
                 $7 \times 7 = 49 + 10 =$ ⑤⑨

                                5925 Ans.

4.) 326   First multiply 26 x 24 $4 \times 6 = 2$④
   x 924 $4 + 6 = 10 + 2 = 1$②
                                            $2 \times 2 = 4 + 2 =$ ⑥
     624       $4 \times 3 + 9 \times 6 = 54$
      66       $9 \times 2 + 2 \times 3 = 24$ 624
      24       $9 \times 3 \qquad = 27$
 +   27
   301224   Ans.

22. Numbers Whose Tens and Units Are Alike

Rule: Square the tens and units. Cross-multiply units and hundreds and add to square two places to left. Add the hundreds digit. Add to above (one place to left). Multiply the hundreds and add to above (one place to left).

1.) 412 x 212
        144   12 x 12
         12   (2 x 4) + (2 x 2)
          6   4 + 2 = 6
 +   8   4 x 2 = 8
     87344 Ans.

2.) 713
   x 613
 4 3 7 0 6 9 ——— $13^2 = 169$. Set down 69, carry 1
               —— $3 \times 7 = 21 + 3 \times 6 = 18$. $21 + 18 + 1(carry) = 40$
               Set down 0, carry 4
               —— $7 + 6 = 13 + 4 (carry) = 17$
               Set down 7, carry 1
               —— $6 \times 7 = 42 + 1 (carry) = 43$

The above can be done mentally. Start with the square and work to the left. You may desire to indicate the carries by a small notation

                4 3 7 0 6 9   Ans.

34

23. Three-Digit Numbers with Same First Two Digits
(Hundreds Digit = 1)

Rule: Add the units digit of one number to the whole of the other number. Multiply this by the second digit. Annex a zero and add to this the product of the units digit.

1.)   121   121 + 8 = 129 +0
   x   128   129 x 2 = 2580
               8 x 1 = +   8
                     15488   Ans.

2.)   192   192 + 8 = 200 +0
   x 198   200 x 9 = 18000
               8 x 2 = +   16
    Ans.   38016

3.)   109   109 + 2 = 111 +0
  x 102   111 x 0 = 0000
            9 x 2 = +   18
                11118   Ans.

The reason 1800 was added without a place over is that it is a 4 digit number.

Another Method

Rule: Square the tens and hundreds digits. Multiply the units digits (product must be two digits). Set down to right. Cross-multiply the tens and units digits and set down product one place to left. Cross-multiply units and hundreds digits and set down one place to the left.
Add for the answer.

1.)   121   12 x 12 = 144
   x 128   1 x 8 = 08
  14408
     18   (18 x 2)+(1 x 2)
 + 09   (1 x 1)+(1 x 8)
  15488 Ans.

2.)   109   10 x 10 = 100
   x 102   9 x 2 = 18
  10018   (2 x 0) + (9 x) = 00
    00   (2 x 1) + (9 x 1) =11
 + 11
  11118 Ans.

24. Three-Digit Numbers with Same First Two Digits
(Hundreds Digit Greater Than 1)

Rule: Add the units digit of one number to the whole of the other number. Do not set down. Multiply this by the hundreds digit. Annex a zero. Now multiply the number in the first step by the tens digit. Add to product above. Multiply the units digits and add to the partial products as shown in example.

```
1.)   216      216 + 2 = 218
   x  212                218 x 2   =  4360      +0
                         218 x 1   = + 218
                                      4578
                          2 x 6     = +    12
                                       45792    Ans.


2.)   439      439 + 3 = 442
   x  433                442 x 4    = 17680    +0
                         442 x 3    = +1326
                                      19006
                          9 x 3     =+      27
                                       190087   Ans.

3.)   37.2     372 + 8 = 380
   x  3.78               380 x 3   = 11400    +0
                         380 x 7   =+ 2660
                                     14060
                          8 x 2    =        16
                                      140616
                Count off 3 places    140.616   Ans.
```

Note: The alternate method shown in 23 can also be used for multiplying when the hundreds digit is more than 1.

25. Four-Digit Numbers Whose First Three Digits Are Alike (An Extra Step Is Involved)

Rule: Add the units digit of one number to the other whole number. Multiply this by 10, 20, 30, etc. (whatever the thousands and hundreds units fall between). Set down. Multiply by hundreds unit. Add to the above (one place to the right). Multiply by tens digit and add to above (one place to the right). Multiply by units digit and add, as shown in examples.

```
1.)   1428                 1434 x 10  =  143400    + 0
   x  1426                 1434 x 4   =  5736

                           1434 x 2   =  2868
   1428 + 6  = 1434          6 x 8    =+    48
   14 is between 10 and 19                2036328   Ans.
   Multiply by 10. Annex a 0.
```

2.) 3326   3328 x 30 = 998400   +0
   x 3322   3328 x 3 = 9984

   3326 + 2 = 3328   3328 x 2 = 6656
   33 is between 30 and 39   2 x 6 = ✝   12
   Multiply by 30. Annex a 0   11048972   Ans.

3.) 7742   7751 x 70 = 5425700   +0
   x 7749   7751 x 7 = 54257

   7742 + 9 = 7751   7751 x 4 = 31004
   77 is between 70 and 79   9 x 2 = +   18
   Multiply by 70. Annex a 0   59992758   Ans.

26. Two Numbers Whose First and Last Digits Are Alike and Whose First or Last Digits Add Up to 10

When units digits are alike and the tens digits add up to 10.

Rule: Square the units. Set down (must be two digits). Multiply the tens. Set down in front of square (to left). Add the unit digits times one hundred.

1.)   93   9 + 1 = 10   2.)   84   8 + 2 = 10
      x 13   3 x 3 =09   x 24   4 x 4 = 16
         909   9 x 1 = 9   1616
      + 300   3 x 100 = 300   + 400   8 x 2 = 16
        1209   Ans.   2016   Ans.   4 x 100 = 400

3.)   312   3 + 7 = 10   4.) 124   1 + 9 = 10
      x 712   12 x 12 = 144   x 924   24 x 24 = 576
      210144   7 x 3 = 21 + 0   90576   9 x 1 + 0 + 90
   +   1200   12 x 100 = 1200   + 2400   24 x 100 = 2400
      222144   Ans.   114576   Ans.

Note: When the hundreds digits are multiplied, a zero must be added to the product. The hundreds and units digits are multiplied by ten and are added to the above one place to the left.

When the tens (and hundreds digits) are alike

Rule: Square the tens (and hundreds) digits. Multiply the tens (and hundreds) and add directly to above.

1.)
```
    73        7 + 3 = 10
  x 77        7² = 49
  4921        7 x 3 = 21
+  700        7 x 100=700
  5621    Ans.
```

2.)
```
    52        8 + 2 = 10
  x 58        5² = 25
  2516        8 x 2 = 16
+  500        5 x 100 = 500
  3016    Ans.
```

3.)
```
   123        7 + 3 = 10
 x 127        12² = 144
 14421        7 x 3 = 21
+ 1200        12 x 100 = 1200
 15621    Ans.
```

4.)
```
    551        9 + 1 = 10
  x 559        55² = 3025
 302509        9 x 1 = 09
+  5500        55 x 100 = 5500
 308009    Ans.
```

When the tens and units digits are alike and the hundreds digits add up to 10.

5.)
```
     444        6 + 4 = 10
   x 644        44 x 44 = 1936
  241936        6 x 4 = 24
+  44000        44 x 1000 = 44000
  285936    Ans.
```

6.)
```
     166        1 + 9 = 10
   x 966        66 x 66 = 4356
  094356        9 x 1 = 09
+  66000        66 x 1000 = 66000
  160356    Ans.
```

Note: The common digit to both is multiplied by 1000.

When the tens and units of each number add up to 100 and the thousand and hundred units are alike.

7.)
```
    3275        25 + 75 = 100
  x 3225        32 x 32 = 1024
 10241875       75 x 25 = 1875
+  320000       32 x 10000 = 320000
 10561875    Ans.
```

8.)
```
    3217        17 + 83 = 100
  x 3283        32 x 32 = 1024
                17 x 83 = 1411
                32 x 10000 = 320000
 10241411 + 320000 = 10561411    Ans.
```

27. Divide and Multiply

Rule: Multiply one of the numbers by some digit to simplify the number and then divide the other number by the same number you used to multiply. Or divide and multiply. This facilitates many multiplications.

1.)   32 x 2 1/4 ($\frac{9}{4}$)
    $\div$ 4   x 4
       8 x 9 ($\frac{9}{4}$ x 4)
       72   Ans.

2.)   49 x 136
    $\div$ 7   x7
       7 x 952
    6664   Ans.

3.)   480 x 480
    $\div$ 60   x 60
       8 x 18800
    230400   Ans.

4.)   40 x 2.75
    $\div$ 4   x 4
       10 x 11
       110   Ans.

5.)   324 x 81
       x 9 $\div$ 9
    2916 x 9
    26244   Ans.

6.)   .0036 x .0072
    or   36 x 72
       $\div$ 6   x6
       6 x 432
       2592
    count off 8 places to the left
       .00002592   Ans.

7.)   433 x 1.6
    $\div$ 10   x 10
       43.3 x 16
    x 8 $\div$ 8
    346.4 x 2
       692.8   Ans.

28. "Funnel Method"

Rule: Multiply the individual digits and set down their products
successively. Cross-multiply the hundred and the ten digits and then the tens and units digits. Set down as shown in
the examples (in form of a funnel). Cross-multiply the first
and last digits and set down underneath above (as shown).
Add for the answer.

Three Digits

1.)   328
    x 576
151448
  3168
+   68
188928   Ans.

a.) 6 x 8 = 48, 7 x 2 = 14, 5 x 3 = 15
b.) (3 x 7) + (5 x 2) = 31
    (7 x 8) + (6 x 2)   = 68
c.) (5 x 8) + (6 x 3)   = 58

Four Digits

2.)   1234   a.) 8 x 4 = 32, 7 x 3 = 21, 6 x 2 = 12, 5 x 1 =05

    X 5678   b.) (7 x 4) + (8 x 3) = 53

                  (6 x 3) + (7 x 2) = 32   05122132

                  (5 x 2) + (6 x 1) = 16   163252

            c.) (5 x 3) + (7 x 1) = 22   2240

                  (6 x 4) + (8 x 2) = 40 +   28

            d.) (5 x 4) + (8 x 1) = 28   70006652   Ans.

Three Digits by Two Digits

3.)   756   a.) 6 x 8 = 48, 7 x 5 = 35

   x   78   b.) (7 x 6) + (8 x 5) = 82   3548

             c.) (7 x 7 = 49   +   82

             d.) 7 x 8 = 56   4368

                                     49

                            +   56

                            58968   Ans.

29. Application of the Funnel Method

Rule: Draw a vertical line after the second place and work out the Funnel Method from left to the third place after the decimal point, rounding off if necessary any numbers to the right of the vertical line.

Example:

```
        a b c d e          A.)  (a x f)        (b x g)
        6. 3 4|5 7               6 x 5 = 30   3 x 6 = 18
        5. 6 7|2 8               (c x h)
        f  g h|i j               4 x 7 = 28 , round off to 30
A.)    30 1 8 |3                              Set down 3
B.)     5 1 4 |5          B.)  (f x b)
C.)       6 2 |4               5 x 3 = 15
D.)         3 |8          plus (g x a)
E.)         _ |8               6 x 6 = 36
       35.9 9  8   Ans.                  51   set down
                                (g x c)
correct to two places       plus 6 x 4 = 24
                                (h x b)
                                7 x 3 = 21
                                         45   set down
```

B.) 6 3 4 |5 7
5 6 7 |2 8

C.) 6 3 4 |5 7
5 6 7 |2 8

D.) 6 3 4 |5 7
5 6 7 |2 8

E.) 6 3 4 |5 7
5 6 7 |2 8

This procedure may look very complicated and tedious but it is very simple once it is learned.

c.) (f x c)
   5 x 4 = 20
+ (h x a)
   7 x 6 = $\underline{42}$
        62   set down
   (i x b)
   2 x 3) = 6   round off to 40
   (g x d)   Set down 4 to
   6 x 5 = $\underline{30}$ right of V line
        36

D.) (f x d)
   5 x 5 = 25
+ (i x a)
   2 x 6 = $\underline{12}$
        37
   (g x e)
   6 x 7 = 42
   (j x b)
   8 x 3 = $\underline{24}$
        46 or 7 0 or .7
   Add to 37 above a .7 Set
        down 38
E.) (j x a) = 48
   8 x 6
+ (f x e) = $\underline{35}$
        8 3 Round off to 8

30. By Numbers Which Are Nearly Whole Numbers

Rule: Multiply by the next highest whole integer. Subtract the part or fraction you increased the multiplier.

1.)   720 x 7 3/4
  or 720 x 7.75

    720 x 8 = 5760
-1/4 of 720 $\underline{- 180}$
        5580   Ans.

2.) 34 x 6 2/3
  or 34 x 6.667

     34 x 7 = 238
-1/3 of 34 = $\underline{11.33}$
      226.66   Ans.

3.) 124 x 7.8
   124 x 8 = 992
  -.2 of 124 $\underline{- 24.8}$
       967.2   Ans.

4.) 7536 x 19.7
   7536 x 20 = 150720
  -.3 x 7536 = $\underline{- 2260.8}$
       148459.2

If you decrease the multiplier to obtain a single whole integer, add the fraction to the semi-product.

```
5.)  75 x 8 1/5              6.)  362 x 121 1/4
     75 x 8.2                     362 x 121.25
     75 x 8  =  600           362 x 100 = 36200
   + 1/5 of 75 +  15          362 x  20 =  7240
               615    Ans.    362 x   1 =   362
                            + 1/4 of 362 = +  90.5
                                         43892.5


7)  45 x 33.875
    45 x   34                 45 x 30 = 1350
                             45 x  4 =  180
                                     + ____
                                        1530
                           -1/8 of 45 -  5.625
                                      1524. 375
```

31. Different Pairs Of Numbers
(Variation Of Cross-Multiplication)

Rule: Multiply the units digits. Double the answer and set down underneath first product, one place to the left. Multiply the tens digits and set down one place to the left. Add all three for answer.

```
1.)  44       12    3 x 4 = 12    2.)  55       20    5 x 4
   x 33       24    double 12       x 44       40    double
   +    12          3 x 4 = 12              + 20
      1452   Ans.                            2420   Ans.
```

When three digits are multiplied by either two or three digits, multiply by the units digit and add it to itself two times, moving one place to the left each time.

```
3.)  111      11 x 8 = 88      4.)  222      22 x 3 = 66
   x  88          88              x 33           66
                  88                             66
              +   88                         +  66
                9768                           7326    Ans.


5.)  999                           Alternate method
   x  77     99 x 7 = 693
              693              6.)  44       3 x 44 = 132
              693                 x 33          132
            + 693                            +  132
             76923   Ans.                      1452
```

42

32. Multiplication by 5

Rule: Divide number by two and multiply it by ten; that is, halve it and append a zero.

1.) 126 x 5
   (÷ 2 x10)
     63 x 10 = 630 Ans.

2.) 458 x 5
     229 x 10 = 2290 Ans.

When number is odd, take one half of number (disregarding the fraction) and append a five.

3.) 167 x 5   1/2 of 167 = 83 + 5 = 835   Ans.

4.) 573 x 5   1/2 of 573 = 286 + 5 = 2865   Ans.

5.) 4533 x 5 1/2 of 4533 = 2266 + 5 = 22665   Ans.

Group Version

Rule: To multiply a number by 5, halve the number up to the tens digit and append the units product (units × five) as a two digit number.

6.) 5 x 1847 $\frac{184}{2}$ = 92 + (7 x 5) = 9235   Ans.

7.) 5 x 33345 $\frac{3334}{2}$ = 1667 + (5 x 5) = 166725   Ans.

8.) 5 x 52300 $\frac{52300}{2}$ = 2615 + 5 x 0 + 0 = 261500   Ans.

9.) 5 x 32121 $\frac{3212}{2}$ = 1606 + 05 (1 x 5) = 160605   Ans.

33. Multiplication by 25

Rule: To multiply a number by 25, multiply a quarter of it by 100, (append two zeros). Then multiply remainder by 25.

1.) 16 x 25   1/4 of 16 = 4, 4 x 100 = 400   Ans.

2.) 19 x 25   1/4 of 19 = 4, plus 3 remainder
                    4 x 100 + 3 x 25 = 475   Ans.

3.) 217 x 25 $\frac{217}{4}$ = 54, plus 1 remainder
                    (54 x 100 = 5400 + 1 x 25) = 5425   Ans.

4.) 344 x 2.5 $\frac{344}{4}$ = 86 + 0 (no remainder), 86 x 100 = 8600
                    count off one place   = 860   Ans.

5.) 344 x 250 $\frac{344}{4}$ = 86 + 0 (from 250) = 860
                              860 x 100 = 86000   Ans.

6.) 464 x 325 $\frac{464}{4}$ = 116 x 100 = 11600 0   (3 x 464)   from
                  (3 x 464) from 325 = 1392   3 25
                                       150800   Ans.

7.) 5673 x 425   1/4 of 5673 = 1418 with 1 remainder
     1418 + 00 =141800 + (1 x 25)   =   141825
                  4 x 5673   +   = 22692
                                       2411025   Ans.

8.) 346 x 252 (units digit is 2)
     $\frac{346}{4}$ = 86 with 2 remainder
     Upend 86 with two zeros   = 8600
     Add 25 x 2(the remainder)   50
     346 x 2 (units digit)   +   692
                                             87192   Ans.

Note: when the odd number is the hundreds digit the product of
this number and the multiplicand is set down one place in
front of the multiplicand. When it is the units digit it is
placed one place to the last digit of the multiplicand.

34. Multiplication by 75

Rule: To multiply a number by 75, multiply a quarter of it by 300
and the remainder by 75.

1.)   48 x 75 $\frac{48}{4}$ = 12 x 300 = 3600 Ans.

2.)   17 x 75 $\frac{14}{4}$ = 4 with 1 remainder
                        4 x 300   = 1200
                        1 x 75   = + 75
                                     1275   Ans.

3.) 326 x 75 $\frac{326}{4}$ = 81 with remainder 2
                       81 x 300 = 24300
                        2 x 75 = + 150
                                    24450   Ans.

4.) 978 x 75 $\frac{978}{4}$ = 244 with remainder 2
                      244 x 300 = 73200
                        2 x 75 = + 150
                                    73350   Ans.

Alternate method for numbers with remainders.

Take the next highest number, 980, which leaves no remainder when divided by 4. Multiply by 300. Then subtract 2×75 or 150 from product. (75 times the amount number was raised).

$$\frac{980}{4} = 245$$

$$245 \times 300 = 73500$$
$$-(2 \times 75) = -\underline{150}$$
$$73350 \quad Ans,$$

35. Multiplication by 125

Rule: Multiply an eighth of the number by 1000 (append three zeros). Multiply remainder by 125.

1.) 48×125 1/8 of $48 = 6$ $\times 1000 = 6000$
125×0 (remainder) $= +\underline{000}$
6000 Ans.

2.) 124×125 $\frac{124}{8} = 15$ with remainder 4
append 000 to 15 $= 15000$
4×125 $= +\underline{500}$
15500 Ans.

3.) 2233×125 $\frac{2233}{8} = 279$ with remainder 1
279000
1×125 $= +\underline{125}$
279125 Ans.

4.) 456×3125 $\frac{456}{8} = 57$ with 0 remainder
3 is in front of 57000
125 . Multiply 456 + $\underline{1368}$
by 3= 1368 1425000 Ans.

5.) 456×1253 $\frac{456}{8} = 57$ with 0 remainder
3 follows 125 Multiply 456 by 57000
3 and add as shown + $\underline{1368}$
571368 Ans.

6.) 340×1.25 $\frac{340}{8} = 42$ with remainder 4
42000
124×4 $= +\underline{500}$
42500
Count off two places 425.00 Ans.

Also holds true for 12.5, .125, etc.

45

36. Multiplication by 15

Rule: To multiply a number by 15, add half of it to itself and multiply sum by 10:

1.) 24 x 15   1/2 of 24 = 12   2.) 78 x 15   1/2 of 78 = 39
         24   78
    + 12   + 39
      36 x 10 = 360   Ans.   117 x 10 = 1170   Ans.

For odd numbers, add to the number the integral part of half of it and append a 5.

3.) 17 x 15   1/2 of 17 = 8.5   Add 8   4.) 59 x 15
      17   59
    + 8   + 29
    255   885   Ans.

5.) 429 x 15   1/2 429 = 214.5
    Add 214 (drop the .5)
      429
    + 214
    6435   Ans.

Or

Use the integral half and multiply by 30 (2 × 15).

6.) 12 x 15 = 6 x 30 = 180   Ans.
7.) 74 x 15 = 37 x 30 = 1110   Ans.
8.) 3756   1/2 of 3756 = 1878
  x 15   1878 x 30 = 56340   Ans.

9.) 4892   1/2 of 4892 = 2446
  x 15   2446 x 30 = 73380   Ans.

for odd numbers

10.) 17 x 15 = 8.5 x 30 = 255   Ans.
11.) 429 x 15 = 214.5 x 30 = 6435   Ans.

Use the integral half, multiply by 30 and add 15 to the answer.

12.) 17 x 15 = 8 x 30 = 250 + 15 = 255   Ans.
13.) 429 x 15 = 274 x 30 = 6420 + 15 = 6435   Ans.

37. Multiplication by 35, 45, and 55

× 35

Rule: Multiply number (multiplicand) by 100. Take one quarter of this number. Add to this one tenth of the number multiplied by 100.

1.) 46   46 x 100 = 4600
     x 35   1/4 of 4600 =   1150
             1/10 of 4600 = + 460
                             1610 Ans.

2.) 758   758 x 100 =   75800
     x 35   1/4 of 75800 = 18950
            1.10 of 75800 = + 7580
                            26530   Ans.

3.) 834   1/4 of 83400 =   20850
     x 335 1/10 of 83400 = +8340
                             29190
                           + 2502
                            279390 Ans.

3 is in front of 335
Multiply 834 by 3 = 2502
and add to 29190

4.) 834   1/4 of 83400 =   20850
     x 353 1/10 of 83400 =+   8340
                               29190
                           +   2502
                              294402 Ans.

3 is in back of 15
3 × 834 = 2502
add as shown

× 45

Rule: Multiply number by 100. Take ½ of this. Subtract ¹/₁₀ of ½ of this number above.

1.) 39   39 x 100 = 3900
     x45   1/2 of 3900 =   1950
            1/10 of 1950 = - 195
                           1755 Ans.

2.) 92   92 x 100 =   9200
     x 45   1/2 of 9200 = 4600
            1/10 of 4600 =- 460
                           4140   Ans.

3.) 756   756 x 100 = 75600
     x 345 1/2 of 75600 =   37800
            1/10 of 37800 = - 3780
                             34020
                         +   2268
                            260820 Ans.

3 is in front of 345
multiply 756 by 3 = 2268
add as shown

47

4.)　756　　　756 x 100 = 75600　　　　　3 follows the 45 in
　　x 453　　　1/2 of 75600 =　37800　　　453.　Multiply 756 by 3
　　　　　　　1/10 of 37800 =- 3780　　　and add as shown.
　　　　　　　　　　　　　 34020
　　　　　　　　　　　+　　2268
　　　　　　　　　　　 342466　Ans.

× 55

Rule: Multiply number by 100. Take ½ of this. To this add ½ of the original number with a zero appended, one place to the right.

1.)　46　　46 x 100 = 4600
　　x 55　　1/2 of 4600 =　2300
　　　　　　1/2 of 46 + 0 =+ 230
　　　　　　　　　　　　 2530　Ans.

2/)　97　　97 x 100 = 9700
　　x 55　　1/2 of 9700 =　4850　　　　When number is odd
　　　　　　1/2 of 97 + 5 = + 485　　　upend ½ of number with
　　　　　　　　　　　　 5335　Ans.　 a 5. Add as shown.

3.)　792　　792 x 100 = 79200
　　x 55　　1/2 of 79200 =　39600
　　　　　　1/2 of 792 + 0 = + 3960
　　　　　　　　　　　　 43560　Ans.

4.)　391　　391 x 100 = 39100
　　x 55　　1/2 of 39100 =　19550
　　　　　　1/2 of 391 + 5 = + 1955
　　　　　　　　　　　　 21505　Ans.

38.　Multiplication by 65, 85, and 95

× 65

Rule: Multiply multiplicand by 100. Take one half. Add to this ¹/₁₀ of the ½ and ¹/₁₀ of the multiplicand × 100.

```
1.)    37          100 x 37 = 3700
    x  65             1/2 of 3700  =    1850
                      1/10 of 1850 =     185
                      1/10 of 3700 = +   370
                                        2405    Ans,

2.)   843          100 x 843 = 84300
    x  65             1/2 of 84300  =   42150
                      1/10 of42150  =    4215
                      1/10 of 84300 = +  8430
                                        54795   Ans.

3.) 4756           100 x 4756 = 475600
    x  65             1/2 of 475600  =  237800
                      1/10 of 237800 =   23780
                      1/10 of 475600 = + 47560
                                         309140   Ans.
```

$$\times\ 85$$

Rule: Multiply multiplicand by 100. Subtract from this $1/10$ of it. Take one half of the number and subtract $1/10$ of this half from the partial product.

```
4.)   32          100 x 32 =   3200
    x  85         - 1/10 of 3200 - 320
                                   2880            2880
                  1/2 of 3200 = 1600            -   160
                  - 1/10 of 1666 == -60            2720    Ans.

5.)  436          100 x 436         =   43600
    x  85         - 1/10            = -  4360
                                        39240
                  1/2 of 43600 = 21800
                       - 1/10 of 21800        -  2180
                                                 37060   Ans.
```

$$\times\ 95$$

Rule: Multiply the multiplicand by 100. Subtract $1/10$ of it. Take one half of number and add $1/10$ of it to the partial product.

```
6.)  76          100 x 76        = 7600
   x 95          - 1/10          = - 760
                                    6840
              1/2 of 7600 = 3800
                 + 1/10 of 3800       +   380
                                         7220   Ans.
```

49

```
7.)  8340          100 x 8340        =   834000
   x   95              - 1/10             - 83400
              1/2 of 834000 = 417000      750600
            + 1/10 of 417000           +   41700
                                          790300  Ans.
```

Multipliers such as 952, 885, 165, etc. can be used. Follow example used for 35.

39. Multiplication by 9 or Its Multiples

Rule: Multiply the multiplicand by the next complete ten, (20, 30, etc.). Subtract from this product $\frac{1}{10}$ of it.

```
1.)  78 x 45      45 = 9 x 5        78 x 50   =   3900
                                   minus 1/10 = -  390
                                                  3510   Ans.

2.) 320 x 72      72 = 9 x 8       320 x 80   =  25600
                                      - 1/10  = - 2560
                                                 23040   Ans.

3.)  56 x 3.6     3.6 = 9 x .4      56 x 4    =    224
                                      - 1/10  = -  22.4
                                                  201.6   Ans.

4.) 9783 x 81     81 = 9 x 9       9783 x 90  =  880470
                                      - 1/10  = - 88047
                                                 792423   Ans.
```

Numbers whose digits are the same are easily multiplied by 9, 18, etc.

```
5.)  77 x 9      63      7 x 9 = 63    6.) 44 x 27     36
           +    63                                +   36
                693                                  396
                                              x 3  ==  1188   Ans.

7.) 666 x 9      54                    8.) 55 x 63         45
                 54                        63 = 9 x 7      45
            +    54                                    +
               5994   Ans.                               495
                                           x7      X      7
                                                        3465   Ans.
```

50

40. Multiplication by 11 or Its Multiples

Rule: To multiply a two-digit number by 11, write the sum of the digits between them.

```
1.)  81 x 11 ,  8 - 1,  8 + 1 = 9        891   Ans.

2.)  67 x 11,   6 - 7,  6 + 7 = 13        6 7
                                        +  13
                                         737
```

Another method. Add the numbers to each other by one place.

```
3.) 67 x 11 =    67          4.) 95 x 11 =   95
              +  67                        +  95
                737  Ans.                  1045   Ans.
```

Longer numbers multiplied by 11.

```
5.)  5413 x 11                     6.)  8749 x 11    Work from
     5    3      5 + 4 = 9              8    9       right to left.
   +  954        4 + 1 = 5          +  623          4 x 9 = 13
    59543        1 = 3 = 4           96239          7 + 4 = 11 + 1 carry
       Ans.                             Ans.        8 + 7 = 15 + 1 carry
```

Multiples of 11, such as 22, 33

```
7.)  13 x 44 = (3 x 4) x 11 = 52 x 11 = 572   Ans.
8.)  62 x 77 = (62 x 7) x 11 = 434 x 11 = 4774   Ans.
```

When 11 occurs as part of multiplier

```
9.)   86  x 115            10.)  4864 x 2511
      946     86 x 11            4864 x 25  =  121600
   +  430     86 x 5            4864 x 11  = + 53504
     9890   Ans.                   *          12213504   Ans
```

*Since you are multiplying four digits by four digits there must be eight places in the answer. So the product of 11 is spaced as shown.

Also note the following examples:

Space and write down the multiplicand 5413 as shown

```
Add 5 + 4 for second term        5 4
Add 1 + 3 for fourth term      +  9143
                                 59543   Ans.
```

12.)   6789 x 11
           Add 6 + 7 for second term   6 7
           Add 8 + 9 for fourth term   +   8 9
                                                  1317
                                                 ────────
                                                  74679   Ans.

13.)   823256 x 11   8 2   5 6
           Add 8 + 2 = 10   3   2
           Add 5 + 6 = 11   +   10 5 11
                                                ────────
           Add 3 + 2 = 5 (middle   9055816   Ans.
                       term)

14.)   743948 x 11   7 4 4 8
           Add 7 + 4 = 11   3 9
           Add 9 +3 = 12   +   111212
                                               ────────
           Add 4 + 8 = 12   8183428   Ans.

By 111

Rule: Split up the number and put the sum of the digits
multiplied by 11 as a two-digit number in the middle.

15.)   65 x 111   6 + 5 = 11 x 11 = 121   6 5
                       work from left   +   121
                                                   ──────
                       to right   7215   Ans.

16.)   89 x 111   8 + 9 = 17 x 11 = 187   8 9
                                                +   187
                                                   ──────
                                                    9879   Ans.

For Larger Numbers

17.)   3456 x 111
           3   6 (4 places in between)
           7   3 + 4   first two places
          12   3 + 4 + 5 first three places
          15   4 + 5 + 6 last three places
          11   5 + 6   last two places
        ────────
        383616   Ans.

18.)   4762 x 111   Work from right tp left

Ans.   5 2 3 8 2 ──── Write 2, add 2 and 6

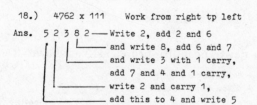

                   and write 8, add 6 and 7
                   and write 3 with 1 carry,
                   add 7 and 4 and 1 carry,
                   write 2 and carry 1,
                   add this to 4 and write 5

52

19.)   2322 x 1111

```
         2     2        2 + 3 = 5
          5            3 + 2 = 5
          5            2 + 2 = 4
   +      4
       25542
   +   225542          Repeat, two places to the right
      2579742   Ans.
```

41. Multiplication by 7½, 12½, etc.
$$\times\ 7\tfrac{1}{2}$$

Rule: 7½ (7.5) is three-quarters of 10. Multiply number by 10 and subtract one-quarter of the product.

```
   1.)  44 x 7 1/2 (7.5)      44 x 10  =  440
                              - 1/4    = -110
                                          330   Ans.
   2.)  392.36 x 7.5      x 10 = 39236
                          - 1/4  - 9809
                                 2942.7   Ans.
```

count off one place

Rule: Multiply number by 10 and add one-quarter of number to product.

```
3.)  44 x 12.5                  4.)  45 x 12.5 (when odd)
     10 x 44        = 440            45 x 10        =   450
     + 1/4 of 440   +110             + 1/4 of 450       + 112.5
                    550   Ans.            Ans.           562.5
```

Or since 12.5 is ⅛ of 100

```
5.)  326 x 12.5
     326 x 100 = 32600,   3260 - 8 = 4075   Ans.

6.)  72 x .125    72/8 = 9   Ans.
```

This method is a variation of times 75 and times 125.

42. Multiplication By 21, 51, 301, 421 (When Last Digit of Multiplier Ends In 1)

Rule: Setting down the product from right to left, write down the units figure of the multiplicand, then multiply each order

53

of the multiplicand by the tens figure of the multiplier, increasing the result in each case by the next highest order of the multiplicand and add any necessary carrying figure.

1.)   278 x 41
        11398   Ans.   Set down the 8.
                                  Multiply 4 x 8 = 32 + 7 = 39
                                  Set down the 9, carry 3.
                                  Multiply 4 x 7 + 2 + 3 (carry) = 33
                                  Set down the 3, carry 3.
                                  Multiply 4 x 2 = 8 + 3 (carry) = 11
                                  Set down the 11.

2.)   3275.75 x 51   Set down the 5 (1 x 5)
        167063.25   Ans.   Multiply 5 x 5 = 27 + 7 =32
                                     Set down the 2, carry 3.
                                     Multiply 5 x 7 = 35 + 5 + 3 carry = 43
                                     Set down the 3, carry 4.
                                     Multiply 5 x 5 = 25 + 7 + 4 carry = 36
                                     Set down the 6, carry 3
                                     Multiply 5 x 7 = 35 + 2 + 3 carry = 40
                                     Set down the 0, carry 4
                                     Multiply 5 x 2 = 10 + 3 + 4 carry = 17
                                     Set down the 7, carry 1
                                     Multiply 5 x 3 = 15 + 1 carry = 16
                                     Set down the 16

3.)   323 x 301   Set down the 3 (3 x 1)
        97223   Ans.   Multiply 30 x 3 (units) + 2 (carry) = 92
                          Set down the 2, carry 9
                          Multiply 30 x 2 = 60 + 3 = 63 + 9 carry = 72
                          Set down the 2, carry 7
                          Multiply 30 x 3 = 90 + 7 carry = 97
                          Set down the 97.

4.)   756 x 421   Set down the 6 (6 x 1)
        318276   Ans,   Multiply 42 x 6 = 252 + 5 = 257
                          Set down the 7, carry 25
                          Multiply 42 x 5 + 210 + 7 = 217 + 25 carry =242
                          Set down the 2, carry 24
                          Multiply 42 x 7 = 294 + 24 carry = 318
                          Set down the 318

Note: When you see a multiplication such as 371×492, use the 371 as the multiplier and follow the procedure as outlined.

5.)   4560 x 1251   Set down the 0 (125 x 0 = 0)
       5704560   Ans.   Multiply 125 x 6 = 750 + 5 + 0 carry = 755
                            Set down the 5, carry 75
                            Multiply 125 x 5 = 625 + 4 = 629 + 75 carry
                            Set down the 4, carry 70   = 704
                            Multiply 125 x 4 = 500 + 70 carry = 570
                            Set down the 570

43. Multiplying Two Numbers Between 11 & 19, 21 & 29, etc.

Rule: Add the unit of one number to the whole of the other. Annex a zero and add the product of the units of both numbers.

        1.) 13 x 19   13 + 9 = 22 + 0   = 220
                                    3 x 9   = + 27
                                                247   Ans.
                    or   19 + 3 = 22 + 0   = 220
                                    3 x 9   = + 27
                                                247   Ans.

        2.)   11 x 15   11 + 5 = 16 + 0   = 160
                                    1 x 5   = + 5
                                                165   Ans.

Two numbers between 21 and 29

Rule: Add the unit of one number to the whole of the other. Multiply by 2 and annex a 0. Add the product of the units.

        3.)   23 x 29   23 + 9 = 32 x 2 =64 + 0 = 640
                                    3 x 9   = +27
                                                      667   Ans.
              For 32 x 38, 91 x 97, etc.

Multiply by the tens digit

        Multiply by the tens digit
        4.)   101 x 109   101 + 9 = 110 x 10 + 0 = 11000
                                    1 x 9   = +   9
                                                     11009   Ans.

55

44. Multiplication by Fractions

Rule: Multiply the multiplicand by the whole digit or digits in multiplier. Change fraction to halves, quarters, eighths, etc. and divide the number by these parts. Add to the partial product.

$82 \times {}^{3}/_{4}$ The conventional method is to multiply 82 by 3 and divide by 4.

1.) Change 3/4 to 1/2 and 1/4

$$
\begin{array}{ll}
1/2 \text{ of } 82 = & 41 \\
1/4 \text{ of } 82 =+ & \underline{20.5} \\
& 61.5 \quad \text{Ans.}
\end{array}
$$

2.) 2240 x 7 5/8
   Change 5/8 to 4/8 and 1/8
      4/8 = 1/2

$$
\begin{array}{ll}
2240 \text{ x } 7 = & 15680 \\
1/2 \text{ of } 2240 & 1120 \\
1/8 \text{ of } 2240+ & \underline{280} \\
& 17080 \quad \text{Ans.}
\end{array}
$$

3.) 360 x 6 2/3
   Change 2/3 to 4/6
      4/6 = 1/2

$$
\begin{array}{ll}
360 \text{ x } 6 = & 2160 \\
1/2 \text{ of } 360 = & 180 \\
1/6 \text{ of } 360 = + & \underline{60} \\
& 2400 \quad \text{Ans.}
\end{array}
$$

4.) 330 x 3 1/3

$$
\begin{array}{ll}
330 \text{ x } 3 = & 990 \\
330 \text{ x } 1/3 =+ & \underline{110} \\
& 1100 \quad \text{Ans.}
\end{array}
$$

5.) 492 x 4 2/7

$$
\begin{array}{ll}
492 \text{ x } 4 = & 1968 \\
2/7 \text{ of } 492 = & \underline{+ 140.57} \\
& 2108.57 \quad \text{Ans.}
\end{array}
$$

45. Trachtenberg System: Multiplication by Single Digit Numbers

Rule: Multiply each digit of the number by the multiplier. Then add the tens digit of the first number to the units digit of the other number for the middle term.

1.)   54 x 6   6 x 5 = 30, 6 x 4 = 24
   3 0 2 4   0 + 2 = 2
   3   2 4   Ans.

2.)　　436 x 7　　　6 x 7 = 42, 3 x 7 = 21, 4 x 7 = 28
　　28 21 42　　　8 + 2 = 10, 1 + 4 = 5
　　2 0　5 2
　　30　　52　　Ans.

3.)　　784 x 8　　　4 x 8 = 32, 8 x 8 = 64, 7 x 8 = 56
　　5 6 6 4 3 2　　6 + 6 = 12, 4 + 3 = 7
　　5
　　　1 2　7 2
　　6 2　7 2　　Ans.

Alternate Method

4.)　　54 x 6　　　4 x 6 = 24, 5 x 6 = 30
　　3 0 2 4　　　　　30 + 2 = 32
　　　324　　Ans.

5.)　　798 x 9　　8 x 9 = 72, 9 x 9 = 81, 7 x 9 = 63
　　63 81 72　　　　63 + 8 == 71 ⎫
　　7 1 8 2　　Ans.　1 + 7 = 8　⎬
　　　　　　　　　　　and 2 = 2　⎭

Multiplication By Two-Digit Numbers

Rule: Multiply units of both numbers. Multiply units of one
number by the tens of the other. Add to this the product
of the tens of first and the units of the second number and
any carry. Multiply the tens of each number and add any
carry.

1.)　a b　　c d
　　6 8 x　1 6　　　　b x d
　　6⁴ 8⁴　8　　　　(b x c) + (a x d) + carry
　10　8　8　Ans.　(c x a) + carry
　　　　　　　　　　8 x 6 = 48　 Write down 8, carry 4
　　　　　　　　　　(8 x 1) + (6 x 6) + 4 carry = 8 +36 + 4 = 48
　　　　　　　　　　Write down 8 and carry 4
　　　　　　　　　　(1 x 6) + 4 carry =10.　Write down 1088 **Ans.**

Rule: The last number of the multiplicand is put down as the
right hand figure of the answer. Each successive digit of
the multiplicand is added to the neighbor at the right.
The first number of the multiplicand becomes the left-
hand number of the answer.

57

× 11

1.)   432 x 11
       4752 Ans.

First digit is 2.
Add next two numbers. 3 + 2
= 5
Add the next two numbers. 4
+ 3 = 7
The last number is 4.

2.)   245671 x 11

    a.) 1 x 1 = ① Set down
    b.) 7 + 1 = ⑧ Set down
    c.) 6 + 7 = 1③ Set down, carry 1
    d.) 5 + 6 = 11 + 1 carry = ② Set down 2
    e.) 4 + 5 = 9 + 1 carry = 1⓪ Set down 0
    f.) 2 + 4 = 6 + 1 carry = ⑦ Set down 7
    g.) 2 x 1 = ② Set down   Ans. 2702381

× 3

Rule: For the first figure, subtract from 10 and double. Add 5 if
the digit is odd. For the middle figures, subtract the digit
from 9 and double what you get. Then add half the neighbor. For the last left-hand figure, divide the first digit of the
multiplicand in half. Then subtract 2.

1.)   02588 x 3

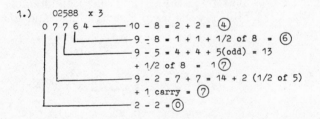

```
0 7 7 6 4 ———— 10 - 8 = 2 + 2 = ④
                9 - 8 = 1 + 1 + 1/2 of 8 = ⑥
                9 - 5 = 4 + 4 + 5(odd) = 13
                + 1/2 of 8 = 1⑦
                9 - 2 = 7 + 7 = 14 + 2 (1/2 of 5)
                + 1 carry = ⑦
                2 - 2 = ⓪
```

× 8

Rule: Subtract units digit from 10 and double. Subtract the tens
digit from 9; double and add the right-hand neighbor. Repeat for the hundreds digit (if necessary). For the first figure of the answer subtract 2 from the first (left-hand) digit
of the multiplicand.

1.)   0789 x 8
6 3 1 2 ——— 10 -9 = 1 x 2(double) = ②
      └——— 9 - 8 = 1 + 1 = 2 + 9(r-h neighbor) = 1 ①
    └———— 9 - 7 = 2 + 2 = 4 + 8 + 1 carry = 1 ③
   └————— 7 - 2 = 5 + 1 carry = ⑥

2.)   0137 x 8
1 0 9 6 ——— 10 -7 = 3 + 3 = ⑥
      └——— 9 - 3 = 6 + 6 + 7 = 1 ⑨
    └———— 9 - 1 = 8 + 8 + 3 + 1 carry = 2 ⓪
   └————— 1 -2 = -1 + 2 carry = (+1)

× 4

Rule: Subtract right-hand digit of the multiplicand from 10, and
add 5 if the digit is odd. Subtract each digit in turn from 9.
Add 5 if digit is odd and add one-half of its neighbor. Under
the zero in front of the multiplicand, write half of its neighbor less one.

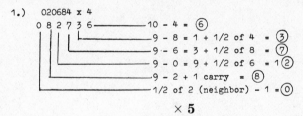

1.)   020684 x 4
0 8 2 7 3 6 ——— 10 - 4 = ⑥
      └——— 9 - 8 = 1 + 1/2 of 4 = ③
     └———— 9 - 6 = 3 + 1/2 of 8 = ⑦
    └————— 9 - 0 = 9 + 1/2 of 6 = 1 ②
   └—————— 9 - 2 + 1 carry = ⑧
  └——————— 1/2 of 2 (neighbor) - 1 = ⓪

× 5

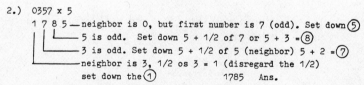

2.)   0357 x 5
1 7 8 5 — neighbor is O, but first number is 7 (odd). Set down ⑤
    └—— 5 is odd. Set down 5 + 1/2 of 7 or 5 + 3 = ⑧
   └——— 3 is odd. Set down 5 + 1/2 of 5 (neighbor) 5 + 2 = ⑦
  └———— neighbor is 3, 1/2 os 3 = 1 (disregard the 1/2)
       set down the ①   1785   Ans.

    Note: When taking ½ of 7 and 9 use 3 and 4.
              When taking ½ of 3 and 5 use 1 and 2

Rule: a.) Subtract the right-hand digit of the number from 10. This gives the right-hand digit of the answer.

b.) Taking each of the following digits in turn, up to the last one, subtract each from 9 and add the neighbor.

c.) At the last step, when you are under the zero in front of the long number, subtract one from the neighbor and use that as the left-hand digit of the answer.

1.) 08769 x 9

   a.) Subtract 9 from 10 to obtain (1.)

   b.) Take 6 from 9.   9 minus 6 = 3

       Add to neighbor. 3 + 9 = 1(2)   Write 2, indicate carry

   c.) 7 from 9 = 2 plus 6(neighbor) = 8 + 1 carry = (9)

   d.) 8 from 9 = 1 plus 7 (neighbor) = (8)

   e.) Last step. Subtract 1 from 8. 8 −1 = (7)

                          Ans.   78921

2.) 01234 x 9

  1 1 1 0 6 ——— 10 − 4 = (6)

            9 − 3 = 6 + 4 = 1 (0)

            9 − 2 = 7 + 3 + 1 = 1 (1)

            9 − 1 = 8 + 2 = 10 + 1 = 1 (1)

            1 − 1 = 0 + 1 (carry) = (1)

Rule: Take the number and double it and add half of the neighbor. Add 5 if number is odd.

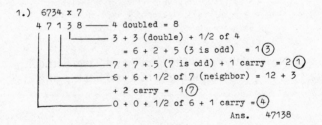

   1.) 6734 x 7

     4 7 1 3 8 ——— 4 doubled = 8

            3 + 3 (double) + 1/2 of 4

            = 6 + 2 + 5 (3 is odd) = 1 (3)

            7 + 7 + .5 (7 is odd) + 1 carry = 2 (1)

            6 + 6 + 1/2 of 7 (neighbor) = 12 + 3

            + 2 carry = 1 (7)

            0 + 0 + 1/2 of 6 + 1 carry = (4)

                       Ans.   47138

× 12

Rule: Double each digit in turn and add to its neighbor.

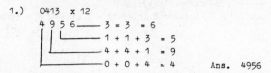

```
1.)    0413  x 12
        4 9 5 6 ──── 3 = 3  = 6
            │   └──── 1 + 1 + 3 = 5
            │  └───── 4 + 4 + 1 = 9
            └──────── 0 + 0 + 4 = 4      Ans.  4956
```

```
2.)    03796  x 12
        4 5 5 5 2 ──── 6 + 6  = 12
            │    └──── 9 + 9 + 6 + 1 carry = 25
            │   └───── 7 + 7 + 9 + 2 carry = 25
            └──────── 3 + 3 + 7 + 2 carry = 15      Ans.  45552
```

× 6

Rule: To each digit add "half" of its neighbor.

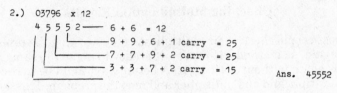

```
1.)    428  x 6
        2 5 6 8 ──────── first number
            │   └──── 2 + 4 (1/2 of 8) = 6
            │  └───── 4 + 1 (1/2 of 2) = 5
            └──────── 0 + 2 (1/2 of 4) = 2      Ans.  2568
```

2.) With odd numbers disregard the fraction or remainder.

```
       0753. x  6
        4 5 1 8 ──── 6 x 3 = 18  Set down the 8, carry 1
            │   └──── 5 + 5 + 1 carry = 11. set down 1   carry 1
            │  └───── 7 + 7 + 1 carry = 15. set down 5   carry 1
            └──────── 0 + 0 + 3 (1/2 of 7) = 3 + 1 carry = 4
                                    Ans.   4518
```

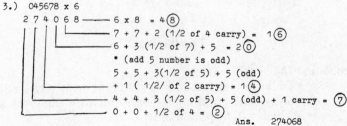

```
3.)    045678 x 6
        2 7 4 0 6 8 ──── 6 x 8 = 4⑧
            │     └──── 7 + 7 + 2 (1/2 of 4 carry) = 1⑥
            │    └───── 6 + 3 (1/2 of 7) + 5 = 2⓪
            │             * (add 5 number is odd)
            │             5 + 5 + 3(1/2 of 5) + 5 (odd)
            │            + 1 ( 1/2/ of 2 carry) = 1④
            │   └─────── 4 + 4 + 3 (1/2 of 5) + 5 (odd) + 1 carry = ⑦
            └────────── 0 + 0 + 1/2 of 4 = ②
                                    Ans.   274068
```

61

*To each "number" add half of the neighbor plus 5 if the number is odd.

```
4.)  034 x 6
     2 0 4 ──────── first number is ④
     │  └───────── 3 + 5 (because 3 is oddd) + 2 (1/2 of 4 neighbor)
     │
     └──────────── 0 + 0 + 1 (1/2 3 nieghbor) + 1 carry = ②  ¹⓪
                                           Ans.  204
```

Checking Multiplication Results

Most people check their multiplication by working the problem out again, as it stands. However, it is easy to make the same mistakes again without noticing it. A better way would be to reverse the multiplicand and multiplier and redo the problem. This is boring and tiresome.

Several easier ways are described and are quicker and more enjoyable. One rule should be abided by and that is to estimate the result before the multiplication is attempted. Round off the numbers to single integers and see how many places are required for the answer. Also, the answer will be approximate and any large discrepancy will be quickly noticed.

Before taking up the "9-Test," a new item will be discussed, CS, the cross-sum of a number. The CS is simply the sum of the integers, repeating the partial sum until a single significant integer is obtained.

The CS (cross-sum) of 63 is 6 + 3 or 9:

```
CS of 95 = 9 + 5 = 14 = 1 + 4 = 5
CS of 147 = 1 + 4 + 7 = 12 = 1 + 2 = 3
CS of 7856 = 7 + 8 + 5 + 6 = 26 = 2 + 6 = 8
CS of 99989 = 9 = 9 = 9 = 8 = 9 = 44 = 4 + 4 = 8
CS of 809070 = 8 + 0 + 9 + 0 + 7 + 0 = 24 = 4 + 2 = 6
```

The "9-Test"

Problem: $734 \times 56 = 41104$

To learn this test it is best to first draw a cross diagonally. The cross is labeled to identify the cross sums which will be placed in it.

62

```
734    x    56   =   41104
```
multiplicand multipler product
Cs   Cb

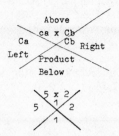

In the left region place the CS
of 734;
$7 + 3 + 4 = 14 = 5$
In the right region place the
CS of 56
$5 + 6 = 11 = 1 + 1 = 2$

Multiply Ca by Cb and enter cross sum of the result in the upper
region. $5 \times 1 = 10 = 1 + 0 = 1$
Take the Cs of the answer and enter it in the lower region. The
upper and lower regions must be equal for the correct answer.

Another example

```
Another example
     498 x 736 = 366528
Ca (498) = 3; Cb (736 ) = 7
Ca x Cb = 7 x 3 = 21 = 3
Cs (366528) = 30 = 3
```

The upper and lower regions
are equal. Answer is correct.

When this check is learned it will not be necessary to draw the
cross. The test can be performed mentally.

Is this test positive all the time?

The answer is no. Errors of position do not show up in the test. Errors in the position of the decimal point are errors of position. Also,
errors of 9 do not show up in the test.

Errors of Position

```
       362              362
    x  308           x  308
       2896             2896
  +    1086      +      10860
       13756            111496
      wrong            right
```

The test shows the answer to be correct but from the example, the
answer is wrong.

To correct for this, estimate the order of magnitude first. 362×308 must have 6 places. (360×300). So the answer 13756 is wrong. Another example for estimating the correct position of the decimal point is illustrated.

$$36.2 \times 14.38 = 520556$$

There must be three places in the answer $(1 + 2)$. Count off 3 places to the left.   520.556   Ans.

Errors of 9

Take 479×201. The 9 in the multiplicand can be typed (or mistaken) for a 0. The 0 in 201 can be typed (or mistaken) for a 9.

$479 \times 201, 470 \times 201, 479 \times 291$ all check out but only the original 479×201 is correct.

```
479 x 201 = 96279     All Ca's = 2
470 x 201 = 94470     All Cb's = 3
479 x 291 = 139389    All answers = 6
```

The CS of a number does not alter if a 9, or any multiple of 9, such as 18, 27, 81, 63, etc., is added.

For example:
```
2 + 9 = 11  Cs = 1 + 1 = 2
3 + 18 = 21  Cs = 2 + 1 = 3
7 + 27 = 34  Cs = 3 + 4 = 7
```

If the test does *not* hold, the answer is *certainly* wrong. If the test is always applied, the number of detected errors will be signficantly increased. Besides, the answer is generally right if the test holds.

Hints on the CS and the nines-test
Cast out any 9-packs first.

```
        7 2 3 6 1  x 2 7 5
    7 and 2, 6 and 3,      2 and 7      left  1 x 5
    · ·  · ·      · · ·
    1 3 5 2 8 1  x  6 2 1 4  =  8 4 0 6 3 6 1 3 4
    5,3 and 1, 8 and 1                      4 x 2
    and 6, 2, 1                              8
                                         2   X   4
                                             8
```

The 9 can be replaced with a 0 when working with CS.

```
                              8 1 9 x 9 7 = 7 9 4 4 3
Remove the 9's                8 1   x   7 = 7 4 4 3
     8 + 1                     9     x   7 = 1 8      (7 +4 + 4 + 3)
Replace 9's with a 0          0     x   7 =  0       (1 + 8 = 9)
```

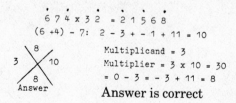

The CS of a number is its *remainder* when the number is divided by 9.

$$\text{CS of } 74 = 2 \qquad 74 \div 9 = 8 \text{ with remainder of } 2$$

The 11-Remainder

Rule: To find the 11-remainder: subtract the *even* sum from the *odd* integer sum

```
ER (eleven remainder) of 18 = 8 - 1 = 7
ER (732) = (7 + 2) - 3 = 6              dots indicate even digits
   . . .                                   (from right to left).
ER (456302) = (2 + 3 + 5) - (0 + 6 + 4)
                11  -  10  = 1   Ans.
```

When the ER is less than 0, add an 11

```
ER (82) = 2 - 8 = - 6            - 6 + 11 = +5
ER (532274) = (4 + 2+ 3) - (7 + 2 + 5)
              odd          even
            = 9 - 14 = - 5       ER = - 5 + 11 = 6
50 ÷ 11 = 4 with remainder of 6
              ER  = 0 - 5 = -5 + 11 = 6
```

The 11-Test

It is the same as the 9-test except that the 11 remainders are used.

```
         6 7 4 x 3 2  = 2 1 5 6 8
         (6 +4) - 7:  2 - 3 + - 1 + 11 = 10
```

```
       8          Multiplicand = 3
   3 X 10         Multiplier = 3 x 10 = 30
       8          = 0 - 3 = - 3 + 11 = 8
    Answer        Answer is correct
```

The 11-test detects errors of position.

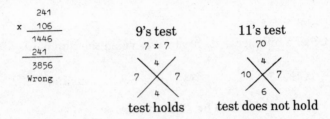

$$\begin{array}{r} 241 \\ \times\ \underline{106} \\ 1446 \\ \underline{241} \\ 3856 \\ \text{Wrong} \end{array}$$

9's test

7 x 7

test holds

11's test

70

test does not hold

Note: There is one exception. If the dividend ER is 0, the dividend is divisible by 11, so the test does not hold. The use of the 9-test and the 11-test together will verify approximately 99% of the multiplications. The 11-test can be used to verify if a number is divisible by 11. If the ER is equal to 0, the number is divisible by 11.

SQUARING

1. Multiplication By Next Highest Digit
(When tens digit is higher than 5)

Rule: Multiply the tens digit by the next highest digit. Square the units digit. Set down. Double units digit and use last integer to multiply the tens digit. Add to the sum obtained, as shown in the examples.

1.) 95^2

$$\begin{array}{r} 9\ 0\ 3\ 6 \\ +\quad 1\ 8 \\ \hline 9\ 2\ 1\ 6 \end{array}$$ Ans.

9 x 10 = 90, 6 x 6 = 36
6 + 6 = 1② 2 x 9 = 18

2.) 78^2

$$\begin{array}{r} 5\ 6\ 6\ 4 \\ +\quad 4\ 2 \\ \hline 6\ 0\ 8\ 4 \end{array}$$ Ans.

7 x 8 = 56, 8 x 8 =64
8 + 8 = 1⑥ 6 x 7 = 42

3.) 105^2

$$\begin{array}{r} 1\ 1\ 0\ 3\ 6 \\ +\quad 2\ 0 \\ \hline 1\ 1\ 2\ 3\ 6 \end{array}$$ Ans.

10 x 11 = 110, 6 x 6 = 36
5 + 6 = 1② 2 x 10 = 20

4.) 127^2

$$\begin{array}{r} 1\ 5\ 6\ 4\ 9 \\ +\quad 4\ 8 \\ \hline 1\ 6\ 1\ 2\ 9 \end{array}$$

12 x 13 = 156, 7 x 7 = 49
7 + 7 = 1④, 4 x 12 = 48

For numbers with units digit less than 5

Rule: Multiply tens digit by next highest digit. Square units digit. Subtract it from 10. Multiply this by tens digit and subtract from partial answer.

1.) 82^2 8 x 9 = 72,   2 x 2 = 04

    7 2 0 4   2 + 2 = 4,   10 - 4 = 6,   6 x 8 = 48

-    4 8

    6 7 2 4   Ans.

2.) 123^2 12 x 13 = 156,   3 x 3 = 09

   1 5 6 0 9   3 + 3 = 6,   10- 6 = 4,   4 x 12 = 48

-    4 8

  1 5 1 2 9   Ans.

When units digit equals 5, multiply and square as before but do not add or subtract anything.

3.) 35^2 3 x 4 = 12,   5 x 5 = 25

   1 2 2 5   Ans.

4.) 95^2 9 x 10 = 90, 5 x 5 = 25

   9 0 2 5   Ans.

5.) 105^2 10 x 11 = 110,   5 x 5 = 25

  1 1 0 2 5   Ans.

2. Algebraic

Rule: Square both digits. Multiply units digit by the tens digit and double the product. Add to the squares as shown in the examples.

1.) 96^5 9 x 9 = 81,   6 x 6 = 36

   8 1 3 6   9 x 6 = 54 x 2 = 108

+ 1 0 8

   9 2 1 6   Ans.

2.) 74^2 7 x 7 = 49,   4 x 4 = 16

   4 9 1 6   7 x 4 = 28 x 2 = 56

+   5 6

   5 4 7 6   Ans.

3.) 111^2 11 x 11 = 121,   1 x 1 = 01

  1 2 1 0 1   11 x 1 = 11 x 2 = 22

+   2 2

  1 2 3 2 1   Ans.

For larger numbers, square each digit separately and set down. Multiply each digit by the other and add as shown. Then multiply the outside digits and add to semi-product as shown in examples.

4.) 798^2

    4 9 8 1 6 4
    1 2 6
+   1 4 4
     1 1 2
  6 3 6 8 0 4 Ans.

7 x 7 = 49, 9 x 9 = 81, 8 x 8 = 64
7 x 9 = 63 x 2 = 126
9 x 8 = 72 x 2 = 144
7 x 8 = 56 x 2 = 112

Now how the 112 is added.

Alternate Method

5.) 798^2

    4 9 8 1 6 4
    1 3 7 2
+   1 2 6 4
  6 4 8 0 0 4
−   1 1 2
  6 3 6 8 0 4 Ans.

7 x 7 = 49, 9 x 9 = 81. 8 x 8 = 64
79 x 8 = 632 x 2 = 1264
7 x 98 = 686 x 2 = 1372
7 x 8 = 56 x 2 = 112

Note: How the partial product are added. Also the outside produce when doubled must be subtracted from the semi-total.

Four-Digit Numbers

6.) 1235^2

    1 4 4 1 2 2 5
+   8 4 0
  1 5 2 5 2 2 5 Ans.

12 x12 = 144, 35 x 35 = 1225
12 x 35 = 420 x 2 = 840

7.) 2513^2

  6 2 5 0 1 6 9
+   6 5 0
  6 3 1 5 1 6 9

25 x 25 = 625, 13 x 13 = 0169 *
25 x 13 = 325 x 2 = 650

*Note: Square must have four digits.

8.) 25.13^2

  631.5169 Ans.

When squaring decimals, set the decimal point to twice the number of places in the original number.

3. Binary Method

Rule: Divide one number by 2 until the digit 1 is reached. Disregard any remainders. Multiply the other number by 2 the same number of times you divided the other number. Cross out the even numbers. Add the odd numbers to obtain the square.

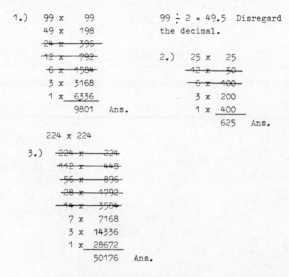

```
1.)   99 x     99          99 ÷ 2 = 49.5  Disregard
      49 x    198          the decimal.
      24 x    396
      12 x    792       2.)   25 x    25
       6 x   1584             12 x    50
       3 x   3168              6 x   400
       1 x   6336              3 x   200
             9801  Ans.        1 x   400
                                     625  Ans.

      224 x 224

3.)  224 x    224
     112 x    448
      56 x    896
      28 x   1792
      14 x   3584
       7 x   7168
       3 x  14336
       1 x  28672
            50176  Ans.
```

4. Napiers Rods

Rule: Draw format with diagonal lines as shown. Square units digit and enter into format as shown. Multiply tens digit by units digit and enter as shown. Square tens digit and enter as shown. Add the diagonal columns. Multiply the units digit by the tens digit and add to partial answer as shown in examples.

```
1.)   7 7 ²              a x a    7 x 7 = 49
       b a               a x b    7 x 7 = 49
                         b x b    7 x 7 = 49
      4/4/4/             b x a    7 x 7 = 49
  + /9/9/9
    5 4 3 9
  +   4 9
    5 9 2 9   Ans.
```

2.)   6 2^2 2 x 2 = 04

      3/1/0/   2 x 6 = 12

+   /6/2/4   6 x 6 = 36

     3 7 2 4   6 x 2 = 12

+   1 2

    3 8 4 4   Ans.

3.)   104^2 4 x 4 =   16

     10/4/1/   10 x 4 =   40

+   /0/0/6   10 x 10 = 100

   10 4 1 6   10 x   4 =   40

+   4 0

   10 8 1 6   Ans.

This method employs the algebraic form but the format is different.

5. Numbers Ending ± 1 of 5 or 10

Rule: Square the number ending in a 5 or 0. Set down the number. Multiply it by 2. Add it to the square if the number was one above or subtract it from the square if number was one below. Add 1 to this number for the final answer.

1.)   36^2 1 above 35   2.)   44^2 1 below 45

         35^2 =   1225   45^2 =   2025

     2 x 35   =+   70   – (45 x 2)   = –   90

                1295   1935

             +   1   +   1

    Ans.   1296   Ans.   1936

3.)   99^2 1 below 100   4.)   201^2 1 above 200

         100^2 = 1000   200^2 =   40000

     –2 x 100   = –200   2 x 200   =+   400

                9800   40401

            +   1   +   1

      Ans.   9801   Ans.   40401

.)   451^2 6.)   349^2

      450^2 =   202500   350^2 =   122500

2 x 450   =   +   900   – 2 x 350= –   700

               203400   121800

            +   1   +   1

       Ans.   203401   Ans.   121801

6. Double and Half Method

Rule: Double one number and half the other. Repeat until one of the numbers reaches a single digit. This method is similar to the binary but differs in that when a single digit is reached you multiply the other number by this single digit. It is easier to multiply by a single digit.

1.) 36^2
     36 x 36
     72 x 18
     144 x 9 = 1296   Ans.

2.) 43^2
     43 x 43
     86 x 21.5
     172 x 10.75
     172 x 10 = 1720
     1/2 of 172 =   86
     1/2 of 86 = + 43
                   1849   Ans.

3.) 17^2
     17 x 17
     34 x 8.5
     34 x 8 = 272
     1/2 of 34 = +17
     Ans.   289

4.) 22 x 22
     44 x 11
     88 x 5.5
     88 x 5 = 440
     1/2 of 88= + 44
               484   Ans.

5.) 104^2
     104 x 104
     208 x 52
     416 x 26
     832 x 13
     832 x 10 = 8320
     832 x 3 = 2496
                10816   Ans.

6.) $.97^2$
     97 x 97
     194 x 48.5
     388 x 24.25
     776 x 12.125
     776 x 10 = 7760
     776 x 2 = 1552
     + 1/8 of 776 =   97
                    94 09

Count off four places .9409 Ans.

7. By Factoring

Rule: When a number can be factored into several single digit numbers, it is better to multiply the number by one factor and the product of this by the other factor (or factors).

1.) 81^2
    9 x 9 = 81
           x   9
           729
       x   9
          6561   Ans.

2.) 49^2
    7 x 7 = 49
           x   7
           343
       x   7
          2401   Ans.

3.) 336^2
    7 x 8 x 6 = 336
             x   7
            2352
        x   8
           18816
        x   6
          112896   Ans.

4.) 241^2 241
    6 x 4 x 10   x   6
    = 240 + 1   1446
    = 241   x   4
                    5784
                x   10
                   57840
              +   241
                   58081   Ans.

5.) 562^2 562
    7 x 8 x 10   x   7
    = 560   3934
    560 + 2   x   8
    = 562   31472
                x   10
                   314720
    +(2 x 562) =   + 1124
                   315844   Ans.

8. Squaring Three-Digit Numbers With Same Digits

Rule: Square the first pair. Square the middle term. Add as shown
in the examples. Note: the middle term forms two identical
pairs with digits at either end.

1.) 666^2 66^2 1st pair   4356
                    3636   2nd pair   4356
    + (36 x 2) + 72   middle term + 36
                    4356   443556   Ans.

2.) 999^2 99^2 9801
                    8181   9801
    81 x 2   +162   +   81
                    9801   998001   Ans.

3.) 3.33^2 33^2
         $3^2 = 09$ 909
              $- \underline{18}$
                   1089

1089
1089
$+ \underline{9}$
110889

Count off four places 11.0889
Ans.

9. Using Boundary Digit*
*Boundary digit is the number above the single significant digit boundary.

Rule: Add the boundary digit to the number. Multiply this by the first digit on the left. Square the boundary digit and add to above as shown in examples.

1.) 412^2 412 is 12 above boudary 400.   412
                  Add 12 to 412. $+ \underline{12}$
                                                       424
                  Multiply by 4 (first   X $\underline{4}$
                               digit)   1696
                  Square 12 and add $+ \underline{144}$
                        as shown   169744   Ans.

2.) 309^2 309 is 9 above boundary   309
                  of 300. Add 9 to 309 $+ \underline{9}$
                                                 318
                  Multiply by 3   x $\underline{3}$
                  (first digit)   954
                  Square 9 and add $\underline{081}$
                  (as showm)   95481   Ans.

3.) 1024^2 1024 is 24 above   1024
                  boundary 1000 $+ \underline{24}$
                  Add 24   1048
                  Multiply by 10   x $\underline{10}$
                  first two digits   10480
                  Square 24 $+ \underline{576}$
                                              1048576   Ans.

When Number Is Below Boundary

Rule: Subtract the boundary digit from the number. Multiply by next higher value of first digit (or digits). Square the boundary digit and add to above as shown in the examples.

1.) 296^2 296 is 4 below boundary of 300.

  296   Subtract 4 from 296
 - 4   Multiply by 3 (1 higher than 2)
  293   Square boundary digit and add. 4^2 = 16
 x 3   (add a zero in front; there must be 3 significant
  876   digits in square)
  016
  87616   Ans.

2.) 985^2 15 below boundary of 1000

  985
 - 15   Subtract 15
  970
 x 10   Multiply by 10 (1 higher than 9)
 9700
+ 225   Square 15 and add as shown
 970225   Ans.

3.) 11989^2 11 below boundary 12000
 - 11   Subtract 11
  11978
x 10 10   Multiply by 12 (Ist a0 x 11978, then 11978
 119780   by 2 and add)
+ 23956
 143736
+   121   Square boundary 11 and add as shown
 143736121   Ans.

10. Written Method for Two-Digit Numbers

Rule: Split the number into tens and units and form the square of each. Double one digit (tens or units) and multiply by the other digit. Write the product under units digit of one and tens digit of the other square. Add all these together. (If one of the squares is a single digit square, write it as a two digit square).

1.) 46^2
 4 | 6
 16 36
+ 48   (4 x 6 x 2)
 21 16

2.) 253^2
 25|3
 625 09
+ 15 0   (25 x 3 x 2)
 640 09   Ans.

77

3.) 28^2
       2|8   (2 x 8 x 2)
       04 64
   +   3 2
       784   Ans.

4.) 1213^2
       12 |13
       1440169
   +   312   (12 x 13 x 2)
       1471369   Ans.

Note: Separate the two squares with 0.

Note: This is another form of the algebraic method.

5.) 2415^2
       24|15
       5760225
   +   720   (24 x 15 x 2)
       5832225   Ans.

When there are 4 digits to be squared a 0 must separate the squares.

11. "Funnel" Method for Numbers with Three or More Digits

Rule: Write in the first row in succession the square of each digit as a two digit number (left to right). Now form the double product of the first digit with every other digit and add as shown in examples.

1.) 642^2
       3 6 1 6 0 4   squares of each digit
         4 8 1 6   6 x 4 x 2 = 48, 4 x 2 x 2 = 16
   +   2 4   6 x 2 x 2 = 24
       4 1 2 1 6 4   Ans.

2.)   a b c d   line 1, squares of each digit
       $(4 3 2 4)^2$ a^2, b^2, c^2, d^2
                        line 2, (a x b x 2) + (b x c x 2) + (c x d x2)
                                double products
                        line 3, (a x c x 2) + (b x d x 2)
                        line 4, (a x d x 2)

       $4 3 2 4^2$
       16090416   squares of each digit
        241216   4 x 3 x 2, 3 x 2 x 2, 2 x 4 x 2 = 24 + 12 + 16
         1624   4 x 2 x 2 = 16, 3 x 2 x 4 = 24
   +   32   4 x 4 x 2 = 32
       18696976   Ans.

6.) $(12\ 3/4)^2$

$$
\begin{array}{rl}
12 \times 12 &= 144 \\
+ 12 &= 12 \\
+ 1/2\ 12 &= 6 \\
+ 3/4 &=+ \underline{\quad .5625} \\
& 162.5625 \quad \text{Ans.}
\end{array}
$$

7.) $(154\ 3/4)^2$ Write $15\ |\ 4$

$$
\begin{array}{l}
 15^2 \ \ 225\ 16 \quad 4^2 \\
15 \times 4 = 60 \times 2 =+ \ \underline{12\ 0} \\
 237\ 16 \\
\text{add} + \ \underline{\ \ 1\ 54} \\
 238\ 70 \\
\text{add } 1/2 \text{ of } 154 + \ \underline{ 77} \\
\text{add as decimal } (.75)^2 239\ 47.5625 \quad \text{ans.}
\end{array}
$$

$(7 \times 8) + (5 \times 5)$

7.)   For eights

$(24\ 3/8)^2$

$$
\begin{array}{ll}
576 & (24)^2 \quad 24/8 = 3 \times 3(\text{numerator}) = 9 \\
+\underline{\ \ 18} & 9 \times 2 = 18 \\
594\ 9/64 & \text{add } (3/8)^2
\end{array}
$$

or   594.1406   (decimal equivalent of 9/64)

12. Squaring Numbers with Fractions

For ½

Rule: Multiply first digit by next higher digit and add the square of ½.

1.($(6\ 1/2)^2$ $6 \times 7 = 42$

$$
1/2 \times 1/2 = \underline{\ \ 1/4} \\
42\ 1/4 \text{ or } 42.25
$$

2.) $(12\ 1/2)^2$ $12 \times 13 = 156 + 1/4$

$$
= 156\ 1/4 \text{ or } 156.25
$$

For ¼

Rule: Square the whole digit. Add ½ of the whole digit to product. Square the fraction and add to product.

3.) $(6\ 1/4)^2$ $6 \times 6 = 36$

$$
\begin{array}{ll}
\text{plus } 1/2 \text{ of } 6 = \underline{+\ 3} \\
1/4 \times 1/4 3\ 9\ 1/16 \\
1/16 = .0625 \text{or} 39.0625
\end{array}
$$

$$4.) \quad (9\ 1/4)^2 \quad 9 \times 9 = 81$$

$$\text{plus } 1/2 \text{ of } 9 = \underline{\quad 4.5}$$
$$85.5$$
$$+ \qquad \underline{.0625}$$
$$85.5625$$

For ¾

Rule: Square the whole number. Add the number and one half of the number to the square. Square the ¾ and add. Or use the decimal equivalent.

$$5.) \quad (8\ 3/4)^2 \qquad 8 \times 8 = 64$$

$$+ 8 = 8$$
$$3/4 = .75 \qquad + 1/2 \text{ of } 8 = 4$$
$$+ .75^2 \qquad = \underline{+ \quad .5625}$$
$$76.5625 \quad \text{Ans.}$$

13. General Rule for All Squaring
(Mirror-Method)

Rule: Transform the square into a product of two mirror numbers of which one is a complete ten. Add the square of the distance from the mirror.

$$1.) \quad 48^2 \quad (50 \times 46) = 2300 \qquad 50 \text{ is } 2 \text{ above } 48$$
$$+ \quad 2^2 = \underline{\quad 4} \qquad 46 \text{ is } 2 \text{ below } 48$$
$$2304 \quad \text{Ans.}$$

So 50 (a complete 10) must be multiplied by $48 - 2$ or 46.

$$2.) \quad 63^2 \quad (60 \times 66) = 3960$$
$$3^2 = + \underline{\quad 9}$$
$$3969$$

Note: 63 is three above 60. So 60 must be multiplied by $63 + 3$ or 66.

$$3.) \quad 105^2 \quad (100 \times 110) = 11000$$
$$+ 5^2 = \underline{+ \quad 25}$$
$$11025 \quad \text{Ans.}$$

$$4.) \quad 98^2 \quad (100 \times 96) = 9600$$
$$+ 2^2 = + \underline{\quad 4}$$
$$9604 \quad \text{Ans.}$$

$$5.) \quad 407^2 \quad (400 \times 414) = 165600$$
$$+ 7^2 = + \underline{\quad 49}$$
$$165649 \quad \text{Ans.}$$

14. Squares of 0-Mirror Numbers (Numbers With Same Digit on Either Side of a Center 0)

Rule: Square first and last digits (or pairs). Set down with two places between the squares. Double one of the squares and enter product in the two places.

1.) 404^2 $4^2 = 16$, $4^2 = 16$, $16 + 16 = 32$
    16   16
      32
  163216   Ans.

2.) 909^2 $9^2 = 81$ 3.) 13013^2
  81--81   81 x 2   169---169
   162   338   169 x 2
  826281   Ans.   169338169   Ans.

4.) 9009^2 5.) 120012^2
 81----81   144-----144
  162   02880
81162081   Ans.   14402880144   Ans.

Note: When squaring two-digit numbers on either side of zero leave three spaces between squares. When there are two zeros in the center leave four spaces between squares. When squaring two-digit numbers whose squares are three digits long and with two zeros in center, leave five spaces between the squares.

15. Squaring 0-Mirror Numbers When Digits Are Not the Same on Either Side of 0

Rule: Square both numbers on either side of zero. Set down and leave two spaces between the squares. Multiply the first and last digits and double the answer. Enter into the two spaces.

1.) 407^2 2.) 906^2
  16--49   81--36
   56   4 x 7 x 2   108   9 x 6 = 54 x 2 = 108
 165649   820836   Ans.

3.) 1203^2
144--091
72   12 x 3 x 2 = 72
1447209   Ans.

4.) 4013^2
16---169
104   4 x 13 = 52 x 2 = 104
16104169

Note how 3^2 is entered as 09   Note how 4^2 is entered.

5.) 72035^2
5184--1225
5040
5189041225

72^2 = 4904
7 x 7 x 2=+ 28
5184

72 x 35 = 2520
x   2
5040

Note: When squaring two-digit numbers on either side of zero, leave only two spaces between squares. Also, if the last number is a two-digit number, leave three spaces between squares.

16. Neighboring Squares

Rule: Take an easy square that is a neighbor of the number to be squared. Multiply the sum of the neigbor and the easy square by the difference of the number and the easy square. Add to the easy square if number is above. Subtract from easy square if number is below.

1.) 57^2 (55^2 is easier)
3025 55^2
+ 224   (55 + 57) x 2 (57 is 2 above 55)
3249   add since 57 is above 55.

2.) 49^2 (50 is easier to square)
2500 50^2
-    99   (50 + 49) = 99 x 1 x 99 (1 difference)
2401   Ans.   Subtract since 49 is below

3.) 152^2 (150 is easier to square)
22500 150^2
+ 604   150 + 152 x 2 (diff) = 604
23104   Ans.

4.) 1998^2 (2000 is easier to square)
4000000 2000^2
-    7996   2000 + 1998 = 3998 x 2 (diff) = 7996
3992004   Ans.   Subtract since 1998 is below

Imagine squaring the number the ordinary way!

82

17. Squaring by Division and Multiplication

Rule: If the number to be squared can be factored (or divided by a number for easier division and multiplication), divide the number to be squared and multiply the number to be squared by the same digit. Multiply these partial products by each other to obtain the answer.

1.) 21^2 $21 = 3 \times 7$ or $\frac{21}{3} = 7$; $21 \times 3 = 63$

$\frac{21}{7} = 3$; $21 \times 7 = 147$ $7 \times 63 = 441$ Ans.

$3 \times 147 = 441$ Ans.

2.) 72^2 $72 = 9 \times 8$

$\frac{72}{9} = 8$; $72 \times 9 = 648$

$8 \times 648 = 5184$ Ans.

3.) 168^2 168 can be divided by 8

$\frac{168}{8} = 21$; $168 \times 8 = 1344$

$21 \times 1344 = 28224$ Ans.

4.) $.036^2$ use 6 as divider and multiplier

$\frac{36}{6} = 6$; $36 \times 6 = 216$

$6 \times 216 = 1296$

count off 6 places $.001296$ Ans.

5.) 448^2 $\frac{448}{64} = 7$; $448 \times 64 = 28672$

$7 \times 28672 = 200704$ Ans.

Alternate Method

1.) 1213   Consider 1213 as 12 and 13.

   x 1213 $13 \times 13 = 169$. Write down last two

digits and carry the 1.   69

$12 + 12 = 24$. $24 \times 13 = 313$

Write down 13, carry the 3   1369

$12 \times 12 = 144 + 3(\text{carry}) = 147$ 1471369   Ans.

To multiply 1213 by itself the conventional way requires twenty-two digits, five lines of figures and many individual calculations. The chance of error increases proportionally.

83

```
2.)    7412        Consider 7412 as 7400 and 12
     x 7412        12 x 12 = 144.  Write down 44 and carry 1
     54937744      74 + 74 = 148.  148 x 12 _  1480    10 x 148
        Ans.       1776 + 1(carry) = 1777    + 296     2 x 148
                   Write 77 (since there       1776
                        are 4 digits)
                   Carry the 17
                   74 x 74 = 5476 + 17(carry) = 5493.  Write down.
```

```
3.)    7412²       The algebraic method of solv-
     74²   12²     ing the same problem is re-
   5476-144        peated   as   a   comparison.
 +   1776          Square 74 and 12. Leave one
   54937744        space between the squares
```
since there must be eight dig-
its. Multiply 74×12 and
double. 1776.
Add as shown in example.

18. Considering the Number as the Sum of Two Numbers

Rule: Set down the given number twice as if for regular multipli-
caton. Assuming that it is considered to consist of tens and
units, multiply the units by units. Write down the unit
product and carry the tens. Add the two given tens together; multiply this sum by the given units; add the carried figure. Write tens in the result and carry the hundreds.
Multiply tens by tens; add the carried figure and write the
result.

```
1.)    6 7        7 x 7 = 49, write 9 and carry 4
     x 6 7        6 + 6 = 12, 12 x 7 = 84 + 4(carry) = 88
     4 4 8 9  Ans.  Write 8 and carry 8
                  6 x 6 = 36 + 8(carry) = 44.  Write 44
```

```
2.)    1 2 4      4 x 4 = 16, write 6 and carry 1
     x 1 2 4      12 + 12 = 24, 24 x 4 = 96 + 1(carry) = 97
     1 5 3 7 6    Write 7 and carry 9
        Ans.
                  12 x 12 = 144 + 9(carry) =153.  Write 153
```

```
3.)    9 7 6        6 x 6 = 36, write 6, carry 3
     x 9 7 6        97 + 97 = 194 x 6 = 1164 + 3(carry) = 1167
     9 5 2 5 7 6    Write 7, carry 116
              Ans.  97 x 97 = 9409 + 116(carry) = 9525.  Write.
```

Note: 97 can be square easily

```
                          97²
                         8149      9² and 7²
                       +  126      9 x 7 x 2
                         9409
```

19. By Simplification

When a number to be squared has a part of it that is an exact multiple of another (such as 124, 749, etc.).

Rule: When the factors are the last two digits, multiply the number by one factor. Multiply the partial product by the second factor. Add as shown in examples (one place to the right).

```
1.)  749²  49 = 7 x 7        2.)  864²    64 = 8 x 8
     749 x 7 =  5243              864 x 8 =  6912
     5243 x 7 =+ 36701            6912 x 8 = + 55296
             561001   Ans.               746496   Ans.
```

When factors are the first two digits, the semi-product is added differently (one place to the left).

```
3.)  124²   12 = 4 x 3        4.)   56.8²    56 = 7 x 8
     124 x 4 =   496                568 x 8 =   4544
     496 x 3 +  1488                4544 x 7 = 31808
            15376  Ans.                      322624
                            Count off 2 places    3226.24   Ans.
```

Note: Multiply by the higher factor first. Also note that in #1 one of the factors (7) is also the first digit. In #3 one of the factors (4) is also the last digit, etc.

20. Three-Digit Numbers with Same First Two Digits

Rule: Add the units digit to the whole number. Multiply this number by the first digit (like digit). Add the whole number to this product (one place over to right). Append a zero. Add to this the square of the units digit.

1.) 114^2 114 + 4= 118   2.) 334^2 334 + 4 = 338

         118 x 1 =   118   338 x 3 =   1014

        +118   =   <u>1180</u>   +0   + 1014   =   10140   + 0

                  12980 4^2 = +   <u>16</u>

        4^2 = +   <u>16</u>   111556   Ans.

             12996   Ans.

3.) 779^2 779 + 9 = 788

           788 x 7 =   5516

         + 5516   = + <u>55160</u>   + 0

                606760

        + 9^2 = + <u>   81</u>

               606841   Ans.

Three-digit Numbers With Same Last Two Digits

Rule: Same as above with one exception. The number obtained by adding last digit to itself must be added to the sum of the product one place to the left.

        4.) 566^2 566 + 6 = 572

                  572 x 5   =   2860

              + 2860   =   <u>28600</u>   + 0

                        314636

                 + 572   <u>572</u>   ⟵

                       320356   Ans.

        5.) 344^2 344 + 4 = 348

                348 x 3   =   1044

            +1044   =   10440   + 0

           + 4^2 = + <u>   16</u>

                   114856

            + 348   + <u>348</u>   ⟵

                 118336   Ans.

Numbers With Same First And Third Digits

        6.) 676^2 676 + 6 = 682

                682 x 6   =   4092

           + 4092   =   40920   + 0

           + 6^2 = + <u>   36</u>

                   450156

          + 682   + <u>682</u>   ⟵

                 456976   Ans.

21. Numbers with Like Pairs of Digits

Rule: Square the first pair. Set down. Double this pair and add to above (two places to the right). To this add the square again (two places to right). Add for answer.

1.) $(1212)^2$
```
        144
        288
    +   144
      1468944   Ans.
```
Square 12
Double square
Add square again

2.) $(7575)^2$
```
        5625
       11250
    +   5625
      57380625   Ans
```
Square 75
Double square. With extra digit add as shown.
Add square again

When All Digits Are Alike

3.) $(3333)^2$
```
        1089
        2178
    +   1089
      11108889   Ans.
```
Square 13
Double 13^2
Add 13^2

When The Pairs Are Different

4.) $(2413)^2$
```
        576
        624
    +   169
      5822569   Ans.
```
Square 24
Multiply 24 by 13 and double

5.) $(1234)^2$
```
        144
        816
    +  1156
      1522756   Ans.
```
Square 12
$(12 \times 34) \times 1 = 816$
34^2

For Five-Digit Numbers

6.) $(34509)^2$
```
     34509 + 9 = 34518
   * 34518 x 345 = 11908710
     Append two zeros = 1190871000
            Add 9² = +        81
                     1190871081   Ans.
```

```
 * 34518 x 345     34518 x 300 = 10355400
                   34518 x  40 =  1380720
                   34518 x   5 = + 172590
                                 11908710
```

87

The approximate answer can be obtained as follows

$$(345)^2 \qquad \begin{array}{l} 345 \times 5 = 350 \\ 350 \times 34 = 119000 + 0 \\ 5^2 = + \underline{25} \\ \text{Append 4 zeros} \quad 1190250000 \end{array}$$

There must be ten digits in the answer.

The above may seem like a lot of work. But certainly it is much easier than to multiply two five-digit numbers. This would require eight lines of figures plus the carries and shifting of places. Also, errors are easily introduced with long additions.

22. By Addition and Multiplication

Rule: Add the units digit to the whole number. Multiply this number by the tens digit. Append a zero. Square the units digit and add to this number.

$$1.) \quad 17^2 \qquad \begin{array}{l} 17 + 7 = 24 \times 1 = 240 \ (+0) \\ 7^2 = \underline{+49} \\ 289 \quad \text{Ans.} \end{array}$$

$$2.) \quad 36^2 \qquad \begin{array}{l} 36 + 6 = 42 \times 3 = 1260 \ (+0) \\ 6^2 = + \underline{36} \\ 1296 \quad \text{Ans.} \end{array}$$

For Three-Digit Numbers

$$3.) \quad 128^2 \qquad 128 + 8 = 136 \qquad \begin{array}{l} 136 \times 12 = 16320 \ (+0) \\ 8 \times 8 = + \underline{64} \\ 16384 \end{array}$$

Multiply 136 x 12 as follows
136 x 10 = 1360
136 x 2 = $\underline{+272}$
1632

$$4.) \quad 784^2 \qquad 784 + 4 = 788$$

788 x 78
788 x 70 = 55160
788 x 8 = + $\underline{6304}$
61464

$$\begin{array}{l} 788 \times 78 = 614640 \ (+0) \\ + 4^2 + + \underline{16} \\ 614656 \quad \text{Ans.} \end{array}$$

For Four-Digit Numbers

5.) 1412^2 $1412 + 12 = 1424$

```
1424 x 10 =   14240                              19936
1424 x  4 = +  5696
              19936      Append 2 zeros        1993600
                             12²            +      144
                                               1993744   Ans.
```

23. Three-Digit Numbers With 0 In Between

Rule: Square the first and last digits and record with two (or three) spaces in center. Multiply first and last digits and double. Add to squares as shown in examples.

```
1.)   303²                      2.)   808²
     9--09    3²                    64--64   8²
      18    3 x 3 x 2            +   128    8 x8 x 2
     91809   Ans.                   652864  Ans.


3.)   604²                      4.)   13011
    36--16   6² and 4²            • 169---121  13² and 11²
   + 48    6 x 4 x 2             +   286    13 x 11 x 2
    364816  Ans.                    169286121  Ans.
```

*Note: Leave 3 places for 5-digit numbers
More difficult squares require previous knowledge of 2-digit square and 2-digit multiplication

```
5.)     36227²
      1296---729   35² and 27²
   +   1944      36 x 27 x 2
    1297944729   Ans.
```

24. Boundary (alternate Method)

Rule: Square boundary (closest number with one significant figure). Add number to boundary and multiply by the times number exceeds or is below boundary. If number is above boundary, add to square. If it is below boundary, subtract from square.

```
1.)  37²   use 40²              2.)  52²    use 50²
                  1600                            2500
     (37 + 40) x3 ⁻231            (50 + 52) x 2 +204
                  1369  Ans.                     2704   Ans.
```

3.)
3010^2 3000^2 = 9000000
                    + 60100   (3010 + 3000) x 10
                       9060100   Ans.

4.) 747^2 use 750^2 (7 x 8 + 5 x 5) + 00
       (747 + 750) x 3   562300
                              = -   4491
                                   558009   Ans.

5.) $(.397)^2$ use 400
       160000
    - 2391   (400 + 397) x 3
    .157609   Ans. (count off 6 places)

25. Numbers Ending with 25

Rule: Square first number (or numbers). Take one half of first number (or numbers) and multiply by ten. Add to the square, one place to right. Append with 625, $(25)^2$.

1.) 625^2 6^2 =   36   2.) 925^2 9^2 =   81
     (1/2 of 6) x10 = + 30   10 x(1/2 of 9) = + 45
     Append with 625   390625 Ans.   Append with   855625
                                        625   Ans.

3.) 1225^2 12^2 =   144
     (1/2 of 6) x 10   = +   60
        Append with 625   1500625   Ans.

4.) 42.25^2 42^2 =   1764
     (1/2 of 42) x 10   =   + 210
                             17850625   Ans.

Decimal point 4 places to left.

Alternate Method

Rule: Square 25. Write 0625. Square first digit (or digits). Write in front of 0625. Multiply first digit or digits by 5. Append with three zeros. Add the two.

5.) 625^2 25^2 =   0625
     6 x 6 = 36 Write 360625
     6 x 5 = 30 x 10 =+ 30000
                          390625   Ans.

6.) 1225^2 $25^2 = 0625$

          $12 \times 12 = 1444$ Write 1440625

    $12 \times 5 = 60 + 000$ +   60000

                                        1500625   Ans.

26. Numbers Ending with 75

Rule: Square the 75. Square the hundreds digit (or hundred and thousands digits). Place in front of the 75 squared. Multiply the units digit 5 by the hundreds (and thousand) digit. Triple product and add as shown in examples.

1.) 375^2

  95625 75^2 and 3^2

+   45 $3 \times 5 = 15 \times 3$

140625   Ans.

2.) 975^2

  815625 9^2 and 75^2

+   135 $9 \times 5 \times 3 = 135$

950625   Ans.

3.) 1175^2

  1215625 11^2 and 75^2

+   165 $5 \times 11 \times 3$

1380625

4.) 7975^2

  62415625 79^2 and 75^2

+   1185 $79 \times 5 \times 3$

63600625   Ans. 79^2 use 80^2

                6400

              −   158

                6242

              −   1

                6241

> **Note.** $75^2 = (7 \times 8) + 25$
> The unit digit of the second product is placed under the 5 of 5625.

5.) 11275^2

  125445625 112^2 and 75^2

+   1680 $112 \times 5 \times 3$

127125625   Ans.

112^2

  12104 11^2 and 2^2 (04)

+   44 $11 \times 2 \times 2$

  12544

27. Numbers Ending with 15, 35, 45, 55, 65, 85, and 95

Rule: Square the tens and units digits. Square the hundreds (and thousands) digits. Multiply the hundreds (and thousands) digits by the number given in table below and add to the squares.

```
                    n 15    multiply by  3
                    n 35       "    by  7
                    n 45       "    by  9
                    n 55       "    by 11
                    n 65       "    by 13
                    n 85       "    by 17
                    n 95       "    by 19
```

```
1.)  235²                          2.)  745²
     41221    2² and 35²                492025    7² and 45²
+    14       2 x 7                +    63        7 x 9
   55225    Ans.                      555025
```

```
3.)  655²                          4.)  195²
     363025   6 ² and 55²                19025    1² and 95²
+    66       6 x 11               +    19        1 x 19
   429025   Ans.                      38025
```

```
5.)  2115²
     4410225•      21² and 15²
+    63
   4473225
```

*When squaring 15 use four digits for answer. $15^2 =$ 0225

```
6.)  2335²
     5291225      23² and 35²
+    161          23 x 7 (from table)
   5452225    Ans.
```

Numbers ending with 5 are easy to square mentally. 35^2 Multiply 3×4 (next higher number) and add 25.

28. Alternate Method for Numbers Ending with 55, 65, 85, and 95

Rule: Square the tens and units. Multiply the hundreds (or thousands) units by the next higher number. Multiply by the digit listed in the table below and add to the partial answer as shown in examples.

```
    n 55    Multiply by 1
    n 65       "    by 3
    n 85       "    by 7
    n 95       "    by 9
```

n denotes the first digit or digits.

1.) 555^2 5 x 6 and 55^2 2.) 765^2
    303025   first digit   564225   (7 x 8) and 65^2
+   5   n is 5. 5 x 1   +   21   7 x 3 (from table)
Ans. 308025   from table = 5   585225   ans.

3.) 485^2
   207225   (4 x 5) and 85^2 4.) 695^2
+ 28   4 x 7(from table)   429025   (6 x 7) and 95^2
  235225   Ans.   + 54   6 x 9 (from table)
                       483025   Ans.

5.) 1365^2
  1824225   (13 x 14) and 65^2 or 13^2 = 169
+ 39   13 x 3 (from table)   plus 13=   13
  1863225   Ans.   182

29. Numbers Beginning with 5

Rule: Square the 5. Add the units digit to the square. Add the square of the units digit to above.

1.) 52^2 2.) 59^2
   5^2 + 2 = 27 5^2 + 9 = 34
     2^2 = + 04 9^2 = + 81
         2704   Ans.   3481   Ans.

3.) 512^2 4.) 565^2
  -5^2 + 12 = 25 5^2 = 25
         12   +65 = 65
   12^2 = 144 65^2 = + 4225
       262144   Ans.   319225   Ans.

Note how units digit is added to square and how the square is added to above.

5.) 5213^2 • 213^2
   5^2 = 25 2^2 and 13^2 = 04–169
   +213 = 213   (2 x 13)(2) =+ 52
   •213^2 = + 45369   45369
      27175369   Ans.   or
                    213 x 200 = 42600
                    213 x 10 = 2130
                    213 x 3 = + 639
                       45369

30. Squaring Three-Digit Numbers (Alternate Method)

Rule: Square the tens and units digits. Multiply the first and third digits; double and add (two places to left). Double first digit (hundreds) and add (one place to left). Square hundreds digit and add (one place to left).

1.) 313^2

169	13^2
18	3 x 3 x 2
6	3 x 2
+ 9	
97969	Ans.

2.) 815^2

225	15^2
80	8 x 5 x 2
16	8 x 2
+ 64	8^2
664225	Ans.

Four-Digit numbers
(Hundreds and Thousands Digit Below 50)

3.) 1412^2

144	12^2
56	14 x 2 x 2
28	14 x 2
+ 196	14^2
1993744	Ans.

4.) 2013^2

169	13^2
120	20 x 3 x 2
40	20 x 2
+400	20^2
4052169	Ans.

5.) • 60 24^2

576	24^2
480	60 x 4 x 2
120	60 x 2
240	•60 x 4
+ 3500	
36288576	Ans.

*With 4-digit numbers with first two digits above 50, multiply the thousands and hundreds digits by 4 instead of 2.

31. Squaring 111, 222, 333, etc. (Alternate Method)

Rule: Square units digit. Multiply by 2 and add to square (one place to the left). Multiply units digit by 3 and add to partial product (one place to left). Multiply units digit by 2 and add to partial product (one place to left). Square first digit and add (one place to left).

94

1.) 111^2

1	1^1
2	1 x 2
3	1 x 3
2	1 x 2
+ 1	1^2
12321	Ans.

2.) 333^2

9	3^2
18	9 x 2
27	9 x 3
18	9 x 2
+ 9	3^2
110889	Ans.

3.) 777^2

49	7^2
98	49 x 2
147	49 x 3
98	49 x 2
+ 49	7^2
603729	Ans.

4.) 999^2

81	9^2
162	81 x 2
243	81 x 3
162	81 x 2
+ 81	9^2
998001	Ans.

Larger numbers can be squared also. Multiply the square by one or more times ($n^2 \times 4$ or $n^2 \times 5$).

5.) 2222^2

4	2^2
8	4 x 2
12	4 x 3
16	4 x 4
12	4 x 3
8	4 x 2
+ 4	2^2
4937284	Ans.

32. Squares of Numbers from 400-600

Rule: The mean of numbers from 400 to 600 is 500. If number is above 500, subtract 250 from the number and square the amount the number is above the mean. Append the square to the subtracted number. If the number is below the mean, subtract 250 from the number and square the amount it is below the mean. Append it to the answer (when square has three digits).

1.) 513^2

513 is 13 above 500

```
    513
-   250
    263
+   169    13²
 263169    Ans.
```

2.) 487^2

487 is 13 below 500

```
    487
-   250
    237
+   169    13²
 237169    Ans.
```

95

3.) 548^2 4.) 467^2

　548 is 48 above 500   467 is 33 below 500
　　548   467
　－ 250   － 250
　　298   217
　＋ 2304 48^2 ＋ 1089 33^2
　300304   Ans.   218089

Note: When the difference squared results in four digits, the number must be added as shown in above examples, (two places to the right).

33. Square of Numbers from 200-400

Rule: When number is above 300, multiply the difference by 6 and add to 900. Square the difference and add to above as shown in the examples. If the number is below 300, multiply the amount it is below by 6 and subtract from 900. Add the square of the difference to above.

1.) 324^2 (24 above)   2.) 297^2 (3 below)
　　900   900
　＋ 144   24 x 6   － 18   6 x 3
　1044   882
　＋ 576 24^2 ＋ 09 3^2
　104976 Ans.   88209 Ans.

For 4-digit numbers use 9000

3.) 3_212^2 (212 above 3000) 212^2
　　9000   4-144 2^2 and 12^2
　＋ 12720   212 x 60
　102720   ＋ 48   2 x 12 x 2
　＋ 44944 $(212)^2$ 44944
Ans. 10316944   Ans.

4.) 2898^2 (102 below 3000) 102^2
　　90000   10004 10^2 and 2^2
　－ 6120   102 x 60   ＋ 40
　83880   10404
　＋ 10404 102^2
　8398404 Ans.

Note that the number 900 is obtained by squaring the hundreds digit of the mean (300) and multiplying it by 100 ($900 = 3^2 \times 100$.).

34. Squares of Numbers from 600-800

Rule: When number is above the mean, multiply the amount over by 14. Add to 4900. Square the amount over and add to above. If number is below the mean, multiply the amount below by 14 and subtract from 4900. Add the amount below squared to the above (as shown in examples).

```
1.)   702²  (2 above)        2.)   712²   (12 above)
      4900                         4900
        28   2 x 14           +    168    12 x 14
  +     04   2²                    5068
      492804  Ans.           +    144    12²
                                  506944   Ans.


3.)   692²  (8 below)        4.)   678²   (22 below)
      4900                         4900
  -   112   8 x 14           -    308    22 x 14
      4788                         4592
  +    64   8²               +    484    22²
      478864  Ans.                459684  Ans.
```

Note that the first two examples employ a variation of the algebraic system. The last two examples are easier than the algebraic method. This method is easily adapted for 4-digit numbers. 4900 is 7 (of the mean) squared times 100.

Four-Digit Numbers

Rule: The same procedure is used with one exception. 490000 is used instead of 4900.

```
5.)   7014²  (14 above)      6.)   7002²   (2 above)
      490000                       490000
        196   14 x 14               28    14 x 2
  +     196                  +     004    2² *
      49196196   Ans.              49028004   Ans.
```

*Must be three places for the square

97

7.) 6995   (5 below)
     490000

    -   70   14 x 5
     489300

    + ___025 5^2
     48930025   Ans.

8.) 5878^2 (122 below)
     490000

    - 1708   122 x 14
     472920

    + 14884 122^2
     47306884   Ans.

Note that the units digit of the number obtained multiplied by 14 is placed under the 5th place zero of the 490000. The square is so placed that there must be eight digits in the answer.

35. Squares of Numbers from 800-1000

Rule: The same procedure is used as before except that the multiplier is 18 and the number is 8100 ($9^2 \times 100$).

1.) 905^2 (5 over)
     8100

    + __90   5 x 18
     8190

    + __25 5^2
     819025   Ans.

2.) 947^2 (47 over)
     8100

    + _846   47 x 18
     8946

    + _2209 47^2
     896809   Ans.

3.) 892^2 (8 below)
     8100

    - _144   8 x 18
     7956

    + __64 8^2
     795664   Ans.

4.) 815^2 (85 below)
     8100

    - 1530   85 x 18
     6570

    + _7225 $(85)^2$
     664225   Ans.

Numbers with Four-Digits

Use 810000 as the number.
Below is shown a difficult number to square.

5.) 8329^2 (671 below)
     810000

    - 12078   671 x 18
     689220

    + _450241 671^2 *
     69372241   Ans.

671 x 18
671 x 10 = 6710
671 x 8 = _5368_
            12078

*use any one of the methods shown to square 3-digit methods.

This is certainly easier and less subject to errors than the standard method.

98

```
5.)  2385²   (15 below 2400)
     576000   24² x 1000
   -    720   24 x (15 x 2)
     568800
   +    225   15²
     5688225  Ans.
```

Note how 720 is subtracted.

```
6.)  4125²     (125 above 4000)
     1600000    40² x 1000
   +   1000     125 x (4 x 2)
     1700000
   +   15625    (125)²
     17015625  Ans.
```

36. Numbers Between 100 and 10,000 or Higher
(Alternate Method)

Rule: Square the hundreds digit and multiply it by 100. Then double the single hundreds digit and use this as a multiplier. If number is above the single hundreds number, multiply by the difference the number is above by the multiplier (twice the hundreds digit). To this add the square of the difference. If the number is below, subtract the difference number is below (×2). Add the square of the difference, as shown in examples.

```
1.)  813²  (13 above 800)        2.)  617²  (17 above 600)
     6400   8² x 100                  3600   6² x 100
   +  208   13 x (8 x 2)           +  204   17 x (6 x 2)
     6608                             3804
 +    159   13²                    +   289   17²
     660969  Ans.                    380689  Ans.


3.)  794²  (6 below 800)         4.)  986²  (14 below 1000)
     6400   8² x 100                  10000  10² x 100
   -   96   6 x (8 x 2)            -    280   14 x (10 x 2)
     6304                             9720
 +     36   6²                     +   196    14²
     630436  Ans.                    972196  Ans.
```

99

Checking Results Of Squaring Numbers

To check the results of squaring numbers, the 9-test and 11-test are employed. These tests are explained in the multiplication section, (Testing Results Of Multiplication).

9-Test

```
        124 x 124 = 15376
Multiplicand x multiplier = product

Cross-sum, Cs, of multiplicand = 1 + 2 + 4 = 7

Cross-sum, Cs, of multiplier = 1 + 2 + 4 =  7

Cross-sum, Cs, of product = 1 + 5 + 3 + 7 + 6 = 22 = 4
```

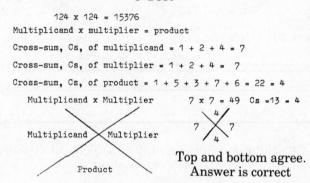

```
 Multiplicand x Multiplier        7 x 7 = 49   Cs =13 = 4
                                        4
 Multiplicand    Multiplier        7    X    7
                                        4
```

Top and bottom agree.
Answer is correct

11-Test

```
   1 2 4 x 1 2 4  = 1 5 3 7 6
```

Subtract sum of even digits from sum of odd digits.
ER (eleven remainder)

```
Multiplicand, (4 + 1) - 2 = 3          3 x 3
                                         9
Multiplier, (4 + 1) - 2 = 3         3        3
                                         9
Product, (6 + 3 + 1) - (7 + 5)
         = 10 -12 = -2
         - 2 + 11 = 9               Test holds
```

CUBING

1. By Factoring (Numbers 11 to 99)

Rule: Cube the tens digit. Set down. Multiply the number by three times the units digit and multiply this product by the tens digit. Cube units digit and add as shown in the examples.

1.) $(13)^3$

1	1^3 (tens digit)
117	$(13 \times 9) \times 1$ (tens digit)
+ 27	3^3 (units digit)
2197	Ans.

2.) $(18)^3$

1	1^3
432	$(18 \times 8 \times 3) \times 1$ (tens digit)
+ 512	8^3 (units digit)
5832	Ans.

3.) $(22)^3$

8	2^3 (tens digit)
264	$(22 \times 2 \times 3) \times 2$ (tens digit)
+ 8	2^3 (units digit)
10648	Ans.

4.) $(57)^3$

125	5^3 (tens digit)
5985	$(57 \times 7 \times 3) \times 5$ (tens digit)
+ 343	7^3 (units digit)
185193	Ans.

5.) $(71)^3$

343	7^3
1491	$(71 \times 1 \times 3) \times 7$ (tens digit)
+ 1	1^3 (units digit)
357911	Ans.

Note how units cube is added. It is easy to remember since the answer must have 6 places. Also the cube must end with 1.

2. Algebraic Method

Rule: Cube the tens digit. Set down. Square the tens digit and multiply by units digit and triple. Add to above (one place to right). Multiply the tens unit by the square of the units digit and triple. Add to above (one place to right). Cube units digit and add (one place to right).

```
1.)  (74)³
            343        7³
            588        (7 x 7 x 4) x 3
            336        ( 7 x 4 x 4) x 3
      +      64        4³ (units digit)
          405224    Ans.

2.)  (39)³
            027        (3)³ (tens digit)
            243        (3 x 3 x 9) x 3  (triple)
            729        (3 x 9 x 9) x 3  (triple)
      +     729        (9)³ (units digit)
           59319    Ans.

3.)  (114)³
           1331       11³ (first two digits)
           1452       (11 x 11 x 4) x 3 (triple)
            528       (11 x 4 x 4) x 3 (triple)
      +      64       4³ (units digit)
          1481544   Ans.
```

3. Multiples of 12

Rule: Cube the tens digit. Multiply the tens and units digits. Multiply this by three; the multiple of the number is of 12. Set down. Double the answer and add. Add the cube of the units digit. Add as shown in the examples.

```
1.)   (48)³           48 = 12 x ④   (48 is 4 x 12).
            64        4³ (tens digit)
           384        (4 x 8) x ④ x 3
           764        (double above)
      +    512        8³ ( units digit)
          110592   Ans.
```

104

2.) $(96)^3$ $96 = 12 \times \boxed{8}$
       729   9^3
      1296 $(9 \times 6)(\boxed{8})(3)$
   - 2592   **(double above)**
   + __216__ 6^3 (units digit)
    884736 Ans.

3.) $(144)^3$ $144 = 12 \times \boxed{12}$
      2744 14^3 (first two digits)
      2016 $(14 \times 4)(\boxed{12} \times 3)$
      4032   (double)
   + __64__ 4^3 (units digit)
    2985984 Ans.

4. Numbers with Same Digits

Rule: Cube tens digit. Product must have three digits. Set down.
Multiply above by 3 and add (one place to right). Add
above again (one place to right). Add the cube of the units
digit (one place to right).

1.) $(66)^3$
       216   6^3 (tens)
       648   216×3
       648   216×3
   + __216__ 6^3 (units)
    287496 . Ans.

2.) $(44)^3$
       064   4^3
       192   64×3
       192   ''   ''
   +   __64__ 4^3
    85184 Ans.

Notice how the 192 is added. The reason is that 4^3 should actually
be written as 064.

3.) $(77)^3$
       343   7^3
      1029 343×3
      1029 343×3
   + __343__ 7^3
    456533 Ans.

5. Multiplying Squares

Rule: Square number using any method outlined previously. Multiply the square by the number for the answer.

1.) $(93)^3$

8109	9^2 and 3^3
+ 54	$(9 \times 3) \times 2$
8649	x 90 = 778410
8649	x 3 = + 25947
	804357 Ans.

2.) $(39)^3$

0981	3^3 and 9^3
+ 54	$(3 \times 9) \times 2$
1521	x 30 = 45630
1521	x 9 = + 13689
	59319 Ans.

3.) $(124)^3$

14416	12^2 and 4^3
+ 96	$(12 \times 4) \times 2$
15376	x 100 = 1537600
15376	x 20 = 307520
15376	x 4 = + 61504
	1906624 Ans.

4.) $(908)^3$

810064	90^2 and 8^2
+ 1440	$(90 \times 8) \times 2$
824464	x 900 = 742017600
824464	x 8 = + 6595712
	748613312 Ans.

6. Boundary Method, Two Digits

When a number is near a boundary which is easily cubed, it is much quicker to cube the boundary and either add or subtract an increment of the number to obtain the answer.

Rule: Cube boundary. Set down. Multiply the number by triple the amount of its difference from the boundary. Append the cube of the units digit. If the number is above boundary, add this to boundary. If number is below boundary, subtract this from the boundary for the answer.

1.) $(99)^3$ 99 is 1 below 100

    1000000 $(100)^3$

−   29701   99 x 3 = 297 Append a 1 as 01 (two places)

    970299   Ans.

2.) $(101)^3$ 101 is 1 above 100

    1000000 $(100)^3$

+   30301   101 x 1 x 3 = 303. Append 1^3 = 01

   1030301   Ans.

3.) $(97)^3$ 97 is 3 below 100

    1000000 $(100)^3$

−   87327   97 x 1 x 3 x 3 = 87327 Append 3^3 = 27

    912673   Ans.

4.) $(105)^3$ 105 is 5 above 100. 5 x 3 = 15

    1000000 100^3

+   157625   (105 x 1 x 15) =   1575

  1157625   Ans. Append 5^3=   125

                            157625

When the Three-digit Front Number Is More Than 1

Rule: Same procedure is employed with one exception. In the second step, when the number is multiplied by three times the difference, the result is multiplied by the hundreds digit of the number before appending the difference cubed.

5.) $(305)^3$

   27000000 $(300)^3$

+ 1372625   300 x 15 = 4575

  28372625   Ans. x 3   x   3

                               13725

             Append 5^3 125

                         1372625

6.) $(297)^3$

   27000000 $(300)^3$

−   801927   297 x 3 x 3 = 2673

  26198073   Ans.   x 3   x   3

                        801927 Append 3^3

This number can be cubed as above but it entails more work.

```
524 x 24 x 3or 72          (500)³  =      125000000
524 x 70 = 36680                             18864
524 x  2 = + 1048      Append 24³   +        13824
           37728                          143877824   Ans.
x 5 (hundreds) x    5
           188640
```

7. Numbers Ending with 5

Rule: Square the number mentally (multiply the tens digit by the next highest digit and append a 25). Append a 0 to the square and multiply by the tens digit. To this, add one half of the appended square for the answer.

1.) $(25)^3$ (2 x 3) +25 = 625. Append a 0.
```
            6250  x 2  =  12500
               + 1/2  = + 3125
                         15625   Ans.
```

2.) $(45)^3$ (4 x 5) +25 = 2025. Append a 0.
```
            20250  x 4  =  81000
            20250 x 1/2 =+ 10125
                           91125   Ans.
```

3.) $(85)^3$ (8 x 9) + 25 = 7225. Append a 0.
```
             72250  x 8  = 578000
             72250 x 1/2 = +36125
                           614125   Ans.
```

4.) $(105)^3$ (10 x 11) + 25 = 11025. Append a 0.
```
             110250  x 10  =  1102500
                 + 1/2  = +     55125
                              1157625   Ans.
```

Note that three-digit numbers are multiplied by 10, 20, etc.

DIVISION

1. By Factoring

Rule: Factor the divisor. Divide number by one divisor. Divide the partial answer by the other divisor.

1.) $4488 \div 24$ $24 = 6 \times 4$

$\dfrac{4488}{4} = 1122$ $\dfrac{1122}{6} = 187$ Ans.

2.) $5000 \div 64$ $64 = 8 \times 8$

$\dfrac{5000}{8} = 625$ $\dfrac{625}{8} = 78.125$ Ans.

3.) $378 \div 56$ $56 = 7 \times 8$

$\dfrac{378}{7} = 54$ $\dfrac{54}{8} = 6.75$ Ans.

Note: If one of the factors of the divisor is an odd number and the other factor an even number, many times it is easier to divide by the odd factor, if the units digit of the dividend or numerator in the fraction is an odd digit, and to divide by even factor if the units digit is an even number.

4.) $1269 \div 24$ $24 = 8 \times 3$

It is easier to divide 1269 by 3 first.

$\dfrac{1269}{3} = 423$ $\dfrac{423}{8} = 52\ 7/8$ Ans.

5.) $262.08 \div 72$ $72 = 9 \times 8$

$\dfrac{262.08}{8} = 32.76$ $\dfrac{32.76}{9} = 3.64$ Ans.

2. By Factoring with More Than Two Factors

Rule: When the divisor can be factored with three or more factors, divide each partial answer by each of the factors in turn.

1.) 46253 ÷ 72 $\frac{46253}{2}$ = 23126.5 $\frac{23126.5}{6}$ = 3854.417
     72 = 6 x 6 x 2
                                        $\frac{3854.417}{6}$ = 642.4 Ans.

2.) 7450692 ÷ 336 $\frac{7450692}{6}$ = 1241782 $\frac{1241782}{7}$ = 177397.428
     336 = 6 x 7 x 8
                          $\frac{177397.428}{8}$ = 22174.678 Ans.

3.) 11.25 ÷ 210 $\frac{11.25}{5}$ = 2.25 $\frac{2.25}{6}$ = .375
     210 = 5 x 6 x 7
                                        $\frac{.375}{7}$ = .05357 Ans.

4.) 1248 ÷ 504 $\frac{1248}{4}$ = 312 $\frac{312}{3}$ = 104
     504 = 3 x 4 x 6 x 7
                                        $\frac{104}{6}$ = 17.33
                                        $\frac{17.33}{7}$ = 2.476 Ans.

3. By Cancellation (Reduction)

Rule: When possible, reduce the dividend and divisor by an
equal amount to obtain a single divisor.

1.) 3114 ÷ 18 is the same as 1557 ÷ 9
                Divide each side by two.

2.) 5211 ÷ 27 is the same as 579 ÷ 3
                Divide each side by 9.

When divisor is even

3.)   5256 ÷ 24 $\frac{5256}{6}$ ÷ $\frac{24}{6}$ 876 ÷ 4 = 219 Ans.

4.)   4256 ÷ 28 $\frac{4256}{7}$ ÷ $\frac{28}{7}$ 608 ÷ 4 = 152 Ans.

When divisor is odd

5.) 2597 ÷ 49 $\frac{2597}{7}$ ÷ $\frac{49}{7}$ 371 ÷ 7 53 Ans.

6.) 176 ÷ 3.3 $\frac{176}{11}$ ÷ $\frac{3.3}{11}$ 16 ÷ .3 = 53.3 Ans.

4. Boundary Division
(Divisor Below Boundary By 1)

(Boundary is the number with only one significant digit)

Rule: Divide number by boundary. Complement the remainder. (By adding the number of times divisor goes into dividend).

```
1.)  82 ÷ 29  Use 30 as the boundary number
     82 ÷ 30 = 2, with remainder 22
     22 + 2 = 24 (complement remainder)
```
$$\text{Ans. } 2 \frac{24}{27}.$$

To express the answer as a decimal, append a zero to the remainder and proceed as previously noted. Add the number of times the boundary goes into each new partial answer.

```
                    82
     30 x 2         60
                    22
       + 2        +  2
                    24.0
     30 x(8)      - 24 0
       ↓            00 0
       + 8          80      8 x 1 complement (30 - 29 = 1)
     30 x(2)      -  60
       ↓            20
       + 2        +  2      2 x 1
                    220
     30 x(7)      - 210
       ↓            10       82 ÷ 29 = 2.827   Ans.
       + 7        +  70  +0
                    170  etc, etc
```

```
2.)  192 ÷ 79          use 80 as boundary
     192 ÷ 80 = 2, remaider 32

      32
    +  2   complement remainder (80 goes into 192 two times)
     340   append zer0
    - 320    ÷ 80 = 4 times
      20   remainder
    +  4   complement
     240   upend zero
    - 240    ÷ 80 = 3 times      Ans.   2.43
       0
```

When dividing larger numbers:

3.) 4563 ÷ 49   use 50
   4563 ÷ 50 goes 91 times
- 4550
    13   remaider
+ 91   complement
  104   ÷ 50 goes 2 times   91 + 2 = 93
+ 100
   40   append a 0   **decimals begin**
+   2   complement
   60   ÷ 50 goes 1 time
- 50
   10   remainder
+   1   complement
  110   ÷ 50 goes 2 times
- 100
   10   remainder
+   2
  120   append 0
- 100   ÷ 50   Ans.   93.122
   20   etc.

5. Boundary Division
(When Number Exceeds Boundary)

Rule: Divide number by the boundary divisor. Subtract and
record remainder. Subtract the complement of the divisor
from the remainder (number of times divisor went into
dividend). Bring down next digit in the dividend and
repeat.

1.) 756 ÷ 61   use 60 as boundary
- 60   60 x 1
  15   remainder
- 1   complement (1 time into 75)
 146   bring doen the 6
- 120   60 x 2
  26   remainder
- 2   complement
 240   upend a 0, decimal starts
- 180   60 x 3 (do not use 60 x 4 since you will get a 0
remainder 60   remainer abd you cannot subtract from 0.
-   3   complement
 570   upend a 0
- 540   9 x 60
  30   etc, etc   Ans.   12.39

114

When divisor is several or more digits above boundary, multiply the number of times boundary goes into the dividend and multiply by how many times divisor is above boundary and subtract this complement from the remainder.

2.)
```
    456  ÷ 52
 -  400    50 x 8
    56
 -   16   remainder
    408   (8 x -2)
 -  350   50 x 7
    58
 -   14   -2 x 7 (complement)
    440   upend a 0 Decimal begins
 -  400    50 x 8
    40
 -   16    -2 x 8
    240   upend 0
 -  200     50 x 4
    40
 -    8    12 x 4
    320   upend 0
 -  300     50 x 6
    20   etc,
```

Do not use 9 × 450. You cannot subtract the complement 9 from 6

Do not use 50 × 8

$\frac{52\;|\;-2}{50}$ format

Ans. 87.846

3.)
```
    .324 ÷ 43
 -  .280     40 x 7
    44
 -   21    -3 x 7
    230   upend zero
 -  200     40 x 5
    30
 -   15    =3 x 5
    250   upend zero
 -  200     50 x 4
    50
```

Do not use 40 × 8

$\frac{43\;|\;-3}{40}$ format

Ans. .007534

Note where decimal is placed.

Simplified Method

```
1.)   35659 ÷ 97                                             97 | +3
                                                             ───────
         3|5657                                               100
      -  3|00        100 x 3                                 format
         56
      +   9       3 x 3 complement
         6|55        100 x 6       Subtract first digit
      +    18        6 x
           7|39        100 x 7
      +       21        7 x 3 complement
Decimal      6|00
      +       18        6 x 3 complement
              1|80
      +         3           1 x 3 complement
               8|3                                      Ans.     367.618
```

```
2.)   3765 ÷ 487                                          487   +13
                                                          ─────────
   -   3500          7 x 500                                 500
       265
                                                            format
   +    91           7 x 13 complement
       3560      append 0, decimal begins
   -   3500          7 x 500
         60
   +     91          7 x 13 complement
       1510
   -   1500          3 x 500
         10
   +     39          3 x 13 complement
        490      Round off to 1                        Ans.     7.731
```

Note: When the difference between divisor and boundary is more than 10, an odd answer will be accurate to three decimal places.

6. Alternate Method of Division By Numbers Near a Unit-Boundary of 100

Rule: Divide the whole number by 100 and determine the total remainder. Complement remainder. Divide this by 100. Final answer is sum of these two products and the remainder if there is one. To convert to decimal, proceed as per example.

1.)   46907 ÷ 97   97 | 3   format

469|07 ÷ 100 = 469 with 07 remainder ^{100}

1407   469 x 3, complement

+   07   add 07 (remainder)

14 | 14   divide by 100 $\frac{1414}{100}$ = 14 with 14

+   42   14 x 3, complement   remainder

560   upend with 0

-    5|00   divide by 100   ⑤ with 60 remainder

60   469

+   15   3 x 5, complement   +   14

750   483   whoe number

-    700   7 x 100   ⑦ with 50 remainder

50

+   21   7 x 3, complement

710

700   7 x 100   ⑦ with 10 remainder

10

+   21   7 x 3 complement   Ans. 483.5773

310

300   3 x 100   ③ with 10 remainder

2.)   508 ÷ 33   3.)   33456 ÷ 33

       5 | 08   3 x 334   1002 | 56

5 x 3   15 | 13   08 + (5 x 1)   334

          Ans. 15 $\frac{13}{33}$ 11 | 390 ÷ 33

                            whole   1013 | 363

       | 130 upend 0   answer   270   upend 0

3 | 99 ÷ 33   8 | 264 ÷ 33

       | 310 upend 0   60   upend 0

9 | 297 ÷ 33   1 | 33 ÷ 33

       | 130 upend 0   270 upend 0

3 | 99 ÷ 33   8 | 264 ÷ 33

       31   Ans.   1013.818

Ans.   15.393

When Divisor Are More Than One Digit Away From Boundary

Rule: Same method is employed as above with one exception. As the hundreds (and thousands) digits on left are brought down to be added to remainder, they must be multiplied by more than one (number away from boundary).

117

4.) 7700 ÷ 32

```
              77 | 00                      32 | +4
     77 x 3  231 | 308   77 x 4          100 | 3
            9 | 288   308 ÷ 32
  Ans.    240 | 200   upend 0
              6 | 192 ÷ 32
                |  80   upend 0
              2 |  64 ÷ 32
                | 160   upend 0
              5 | 160
                |  00        Ans.   240.625
```

7. If Multiple of Divisor Is Near A Unit-Boundary

Three-Digit Numbers

Rule: Divide number by 100. Do not record the two places on the right hand side. Multiply the hundreds digit by multiple of divisor. Add to the cut-off digits, the complement-remainder of the left hand side. Divide this by the divisor and set down the remainder. To the multiplied hundreds digit, add the number of times divisor went into remainder. This is the answer, with remainder over divisor as a fraction.

1.) 6387 ÷ 33

```
              63 | 87                    33 | +1
             189 | 63                   100 | x 3
           +  4 | 150        33 x 3 +1 = 100
             193 with 18 remainder    150 ÷ 33 = 4 with 18 remainder
                                      Ans.    193 18/33
```

To convert to decimal, append a zero to remainder and divide by divisor. Append remainder with zero and divide by divisor. Carry out this procedure as many times as desired.

Decimal Conversion

```
remainder | 180  upend zero
         5 | 165 ÷ 33
           | 150  upend zero          Ans.   193.545
         4 | 132 ÷ 33
           | 180  upend zero
         5 | 165 ÷ 33
           |  15 etc,
```

8. Combination of Factor-and-Boundary Division

Rule: Factor divisor when one factor is close to a boundary. Divide the dividend by boundary method. Divide the answer by the single digit factor.

```
1.)   4237 ÷ 87      87 = 29 x 3
      42 37
      30            30 x ①
      12
    +  1    complement
      133   bring down the 3                146.1 ÷ 3
      120        30 x ④                        48.70   Ans.
       13
    +  4    complement
      177   bring down 7           *If 30 × 5 were used you
  *   180        30 x ⑥            would get the following:
    -   3
    +  6    complement                         177
       30   bring down 0 (decimal starts)      150
                 30 x ①                         27
                                           +  5   complement
                                              320
```

the 30 would divide by more
than ten times into 320.

9. Short Division

Rule: Separate the numbers of the dividend (as shown). Write the remainders as small numbers and place the quotient under the number.

```
1.)    376539 ÷ 7                        3 7₂5₅5₆3₀9
       37 ÷ 7 = 5, remainder 2    remainder 2    5 3 7 9 1    2/7
       26 ÷ 7 = 3, remainder 5    and next digit 6
       55 ÷ 7 = 7, remainder 6
       63 ÷ 7 = 9, remainder 0
       09 ÷ 7 = 1, remainder 2

       The decimal can be calculated as follows:
     ν• 0₆ 0₄ 0₅ 0         2.0 ÷ 7 = 2, remainder 6
       • 2  8  5            60  ÷ 7 = 8, remainder 4
                            40  ÷ 7 = 5, remainder 5
                              etc, etc
       Ans.   53791.285
```

119

2.) $457 \div .4$

```
 4 ÷ 4 = 1, remainder 0          4  5, 7,0, 0
                                  o         o
05 ÷ 4 = 1, remainder 1          1  1  4  2    etc,
17 ÷ 4 = 4, remainder 1
10 ÷ 4 = 2, remainder 2
20 ÷ 4 = 5, remainder 0          Ans.   1142.5
```

Note: $457 \div 4$ is same as $4570 \div 4$

10. By the Short Continental Method

Rule: Procedure will be outlined step by step in the example given.

1.)
$$
\begin{array}{r}
21.45 \\
37 \overline{)793.65} \\
53 \\
166 \\
185
\end{array}
$$

Divisor goes into 79 twice.
Write 2 in quotient.
$37 \times 2 = 74$, deduct mentally from 79.
Write down remainder 5 under the 9.
Bring down the 3. $53 \div 37$ goes once.
Write down 1 in quotient.
Subtract 37 from 53 mentally.
Write down the difference, 16, under the 53.
Bring down the 6. Divide 166 by 37 mentally.
Write down 4 in quotient and write difference of 18 under 66.
Bring down the 5 from dividend.
Divide 185 by 37, which is 5.
Write down in quotient. There is no remainder.

2.) $86 \div 27$
$$
\begin{array}{r}
3.18 \\
27 \overline{)86.} \\
50 \\
230 \\
140 \\
135
\end{array}
$$

```
Mentally      3 x 27 = 81, remainder 5
   "         27 x  1 = 27,      "      23
   "         27 x  8 = 216,     "      14
   "         27 x  5 = 135
            etc,
```

This process reduces the amount of pencil-and-paper work and speeds up the computation.

120

11. By Aliquot Parts

Rule: If the divisor is an aliquot part of a single digit number, multiply the number to be divided by the aliquot number of the divisor and divide by the single digit number.

1.) $1840 \div 25$ $100 \div 25 = 4$
     $1840 \times 4 = 7360 \div 100 = 73.6$ Ans.

2.) $76 \div 25$ $75 \times 4 = 304$
     $304 \div 100 = 3.04$ Ans.

3.) $2480 \div 20$ $100 \div 20 = 5$
     $2480 \times 5 = 12400 \div 100 = 124$ Ans.

4.) $3250 \div 33$ $100 \div 33 = 3 + 1$
     $3200 \times 3 = 9600 \div 100 = 96$

For greater accuracy add $1/100$ of answer above
        96.96   Ans.

5.) $7970 \div 66$ 66 is twice 33
     $\dfrac{7970 \times 3}{100} = \dfrac{23910}{100} = 239.1$

        plus 1/100 of 239.1 $+ \underline{\quad 2.391}$
                               241.491

          $241.49 \div 2 = 120.745$ Ans.

6.) $3250 \div 12.5$ 12.5 is 1/8 of 100
     $3250 \times 8 = 26000 \div 100 = 260$ Ans.

12. By Subtraction

Rule: Multiply the divisor by a larger number not to exceed the dividend. Subtract this product from the dividend. Multiply divisor by a number not to exceed the partial quotient. Repeat as necessary. Add these numbers which were used to multiply divisor to obtain the answer.

1.) $7960 \div 12$
       $-\underline{7200}$ 12 x 600   600
         760   60
       $-\underline{720}$ 12 x 60   3.
          40                            $+ \underline{\quad .3}$
       $-\underline{36}$ 12 x 3   663.3   Ans.
    remainder 4
          40   upend 0
       $-\underline{36}$ 12 x 3
           4   etc,

121

2.)   999 ÷ 7   100   3.)   1 ÷ 8
    - 700   7 x 100   40   .100
      299   2.   - . 96   8 x 12
    - 280   7 x 40   .7   . 40
       19   + .01   -   40   8 x 5
    - 14   7 x 2   142.71   Ans.   .125   Ans.
        5 remainder
       50 upend 0
    - 49   7 x 7
       10 upend 0
    -   7   7 x 1
        3
          142 5/7
    or   142.71   Ans.

13. Division by 5

Rule: Double the tens (or hundreds) digit and divide by 5.

1.) 20 ÷ 5   2 x 2 = 4   0 ÷ 5 = 0   Ans. 4

2.) 37 ÷ 5   (3 x 2 = 6)   (7 ÷ 5 = 1$\frac{2}{5}$)   divide units digit by 5

                                    6 + 1$\frac{2}{5}$ = 7$\frac{2}{5}$ or 7.4   Ans.

3.) 326 ÷ 5   32 x 2 = 64
                 6 ÷ 5 = + 1 1/5
                          65 1/5 or 65.2   Ans.

4.) 4269 ÷ 5   462 x 2 =   852
                   9 ÷ 5 = +   1 4/5
                             853 4/5 or 853.8   Ans.

Alternate Method

Rule: Multiply divisor and dividend by a number which will make the divisor with one significant digit to make for easier division.

5.)   185 ÷ 5   2 x 185 = $\frac{370}{10}$ = 37   Ans.
                     2 x 5   =

6.)   133 ÷ 5   2 x 133 = $\frac{266}{10}$ = 26.6   Ans.
                     2 x 5   =

7.) $\frac{4137}{5}$ 2 x 4137 = $\frac{8274}{10}$ = 827.4   Ans.
                     2 x 5

14. Division by 15

Rule: Write the 15 as 3 × 10. Invert this fraction and multiply the number with this inverted fraction:

1.) $500 \div 15$ $500 \times \dfrac{2}{3 \times 10}$ $= \dfrac{1000}{30}$ = 33.33   Ans.

2.) $74 \div 1.5$ Write 1.5 as $\dfrac{3}{2}$

$74 \times \dfrac{2}{3}$ $= \dfrac{148}{3}$ = 49.333   Ans.

Alternate Method

Rule: Multiply the divisor by 2. Multiply the number by 2. Divide number by divisor.

3.) $135 \div 15$ $\begin{aligned} 2 \times 135 &= \\ 2 \times 15 &= \end{aligned}$ $\dfrac{270}{30}$ = 9   Ans.

4.) $1000 \div 15$ $\begin{aligned} 2 \times 1000 &= \\ 2 \times 15 &= \end{aligned}$ $\dfrac{2000}{30}$ = 66.6   Ans.

5.) $.475 \div 1.5$ $\begin{aligned} .475 \times 2 &= \\ 1.5 \times 2 &= \end{aligned}$ $\dfrac{.950}{3}$ =   = .3166   Ans.

15. Division by 25

Rule: Multiply the hundreds (and higher) digits by 4. Divide the tens and units digits by 25.

1.) $758 \div 25$ $7 \times 4 = 28$

$58 \div 25 = \underline{\;\;2\;}$ and 8 remainder

30 8/25   Ans.

$1/25 = .04 \times 8 = .32$ Ans.   30.32

or $8 \times 4 = .32$

2.) $3240 \div 25$ $32 \times 4 = 128$

$40 \div 25 = + \underline{\;\;1\;} \dfrac{15}{25}$

$15 \times 4 = .60$ 129   15/25

129.6   Ans.

3.) $77921 \div 25$ $4 \times 779 = 3116$

$21 \div 25 = + \underline{\;\;0\;}$ remainder 21

3116

$.21 \times 4 = \underline{\quad .84\quad}$

3116.84   Ans.

123

4.) 13 ÷ 25   no hundreds

   13 ÷ 25 = 0 remainder 13

   13 x 4 = .52   0.52 Ans.

5.) .15 ÷ 25   no hundreds

   .15 ÷ 25 = .00 and $\frac{15}{25}$

   .15 x .4 = .006   .006 Ans.

6.) .27 ÷.25   .27 ÷ .25 = 1. and $\frac{2}{25}$

   .2 x .4 = .08   1.08 Ans.

16. By 75 or Multiples of 25

Rule: 75 is a multiple of 25 (25 × 3). Divide number the same way as by 25. Then divide the partial answer by the multiple digit.

1.) 5|20 ÷ 75   5 x 4 =   20
     Hundreds x 4   20 ÷ 25 = + .8
                                     ‾‾‾‾‾
                                     20.8 ÷ 3
     Tens and units ÷ 25   6.933 Ans.

2.) 48732 ÷ 75   487 x 4 = 1948
     487|32   32 ÷ 25 = 1 $\frac{7}{25}$
     $\frac{7}{25}$ = .7 x .4 = .28   = + .28
                                              ‾‾‾‾‾‾
                                       1948.28   ÷ 3
                                        649.76 Ans.

3.) 830652 ÷ 175   175 = 7 x 25
     8306|52   8306 x 4   =   33224
     $\frac{2}{25}$ = .2 x .4 = .08   52 ÷ 25   = +   2.08
                                              ‾‾‾‾‾‾‾
                                        33226.08   ÷ 7
                                         4746.58 Ans.

4.) 2756 ÷ 125   27 x 4 =   108
                         56 ÷ 25 = +   2
     $\frac{6}{25}$ = .6 x .4 = .24   110.24   ÷ 3
                                          22.04 Ans.

5.) 4579 ÷ 150   4 x 45 =   180
                         79 ÷ 25 = +   3
     $\frac{4}{25}$ = .4 x .4   ‾‾‾‾‾
           = .16   183.16   ÷ 6
     (150 ÷ 25 = 6)   30.52 Ans.

17. Division by 125

Rule: Multiply the thousands digit by 8 and divide the rest by 125.

1.)   121098 ÷ 125   Multiply the remainder by 8
x 8   121|098   and write as decimal. 7 × 098
      968.784 Ans.   = .784

2.)   50 ÷ 125   3.)   8 ÷ 125
      50|050   .05 x 8 |008   .008 x 8
      0. 4   =4   .064   = .064   Ans.

      .4 Ans.

4.)   .006 ÷ 125   5.)   425 ÷ 12.5
      Write .006 as   Write as 4250 ÷ 125

      |.000006 x 8   4 |.250   .250
      .000048 or   4 x 8   32   x   8
      4.8×10^{-5} Ans.   + 2   2.000
                                   34 Ans.

6.)   451256 ÷ 1250   Write as 45125.6 ÷ 125
      45|.1256
x 8   360   .1256 x 8 = 1.0048
+   1.0048
      361.0048 Ans/

18. Division by 11

Rule: Divide the first two digits by 11. Divide remainder and next digit by 11. When the end of the whole number is divided, the decimal remainder is multiplied by 9. This can be repeated as many times as desired.

1.)   563 ÷ 11   56 ÷ 11 = 5, remainder 1
      (remainder 1   13 ÷ 11 = 1, remainder 2
      and next digit   2 x 9 = .181818
      3).   Ans.   51.1818

2.)   3034 ÷ 11   30 ÷ 11 = 2, remainder 8
                       83 ÷ 11 = 7, remainder 6
                       64 ÷ 11 = 5, remainder 9
                        9 x 9 = 8181 Ans.   275.8181

3.) $\frac{245.76}{11}$

.(use decimal)

24 ÷ 11 = 2, remainder 2
25 ÷ 11 = 2, remainder 3
37 ÷ 11 = .3, remainder 4
46 ÷ 11 =4, remainder 2
   2 x 9 = 1818

Ans.   22.341818

To divide by multiples of 11 as 22, 33, 44, etc., divide the numbers by 2, 3, 4, etc., and divide remainder by 11.

4.) $\frac{2364}{22}$ = $\frac{1182}{11}$

Divide by 2

11 ÷ 11 = 1, remainder 0
08 ÷ 11 = 0, remainder 8
82 ÷ 11 = 7. remainder 5
   9 x 5 = 45

Ans.   107.45

5.) $\frac{53.46}{33}$ = $\frac{17.82}{11}$

Divide by 3

17 ÷ 11 = 1, remainder 6
6.8 ÷11 =.6, remainder 2
22 ÷ 11 = 2, remainder 0

Ans.   1.62

6.) $\frac{1200}{4.4}$ = $\frac{300}{11}$

Divide by 4

30 ÷ 11 = 2, remainder 8
80 ÷ 11 = 7, remainder 3
30 ÷ 11 = 2, remainder 8
   9 x 8 = .7272

Ans.   272.7272

7.) $\frac{34567}{111}$

Do not forget .4

345 ÷ 111= 3, remainder 12
126 ÷ 111= 1, remainder 15
157 ÷ 111= 1, remainder 46
460 ÷ 111= .4, remainder 16
   16 x 9 = 144

Ans.   311.4144

8.) $\frac{45676}{220}$ = $\frac{22838}{110}$

Divide by 2

22 ÷ 11 = 2, remainder 0
08 ÷ 11 = 0, remainder 8
83 ÷ 11 = 7, remainder 6
68 ÷ 11 = 6, remainder 2
   2 x 9 = .1818 etc,

Ans.   207.1818

9.)   3300 ÷ 125   8 x 3300 = $\frac{26400}{1000}$ = 26.4   Ans.
                      8 x 125   =

                      ÷ 112$\frac{1}{2}$

10.)   1980 ÷ 112$\frac{1}{2}$ 8 x 1980 = $\frac{15840}{900}$ = 17.6   Ans.
                        8 x 112$\frac{1}{2}$ =

11.) $1000 \div 11,25$ $\begin{array}{l} 8 \times 1000 = \\ 8 \times 11.25 = \end{array} \dfrac{8000}{90} = 88.89$ Ans.

12.) $2700 \div 1125$ $\begin{array}{l} 8 \times 2700 = \\ 8 \times 1125 = \end{array} \dfrac{21600}{9000} = 2.4$ Ans.

$\div 37\frac{1}{2}$

13.) $675 \div 37\frac{4}{2}$ $\begin{array}{l} 8 \times 675 = \\ 8 \times 37\frac{1}{2} = \end{array} \dfrac{5400}{300} = 18$ Ans.

14.) $100 \div 3,75$ $\begin{array}{l} 8 \times 100 = \\ 8 \times 3.75 = \end{array} \dfrac{800}{30} = 26.67$ Ans.

15.) $5000 \div 375$ $\begin{array}{l} 8 \times 5000 = \\ 8 \times 375 = \end{array} \dfrac{40000}{3000} = 13.33$ Ans.

$\div 62\frac{1}{2}$

16.) $812.5 \div 62.5$ $\begin{array}{l} 8 \times 812.5 = \\ 8 \times 62.5 = \end{array} \dfrac{6500}{500} = 13$ Ans.

17.) $12.24 \div 62.5$ $\begin{array}{l} 8 \times 12.24 = \\ 8 \times 62.5 = \end{array} \dfrac{97.92}{500} = .195$ Ans.

18.) $7500 \div .625$ $\begin{array}{l} 8 \times 7500 = \\ 8 \times .625 = \end{array} \dfrac{60000}{5} = 12000$ Ans.

19. Division by 5, 15, 7½, 12½

Rule: Multiply the divisor and dividend by a number which will make the divisor with one significant digit to make for easier division.

$\div 5$
1.) $185 \div 5$ $\begin{array}{l} 2 \times 185 = \\ 2 \times 5 = \end{array} \dfrac{370}{10} = 37$ Ans.

2.) $1330 \div 5$ $\begin{array}{l} 2 \times 1330 = \\ 2 \times 5 = \end{array} \dfrac{2660}{10} = 266$ Ans.

$\div 15$

3.) $135 \div 15$ $\begin{array}{l} 2 \times 135 = \\ 2 \times 15 = \end{array} \dfrac{270}{30} = 9$ Ans.

4.) $1000 \div 15$ $\begin{array}{l} 2 \times 1000 = \\ 2 \times 15 = \end{array} \dfrac{2000}{30} = 66.6$ Ans.

5.) .475 ÷ 15   2 x .475 = .950   = .03167 Ans.
                    2 x 15   = 30

                    ÷ 7.5

6.) 390 ÷ 7.5   4 x 390 = 1560   = 52 Ans.
                    4 x 7.5 =   60

7.) .077 ÷ 7.5   4 x .077 = .308   = .0102 Ans.
                    4 x 7.5 = 60

8.) 175 ÷ 12.5   8 x 175 = 1400   = 14 Ans.
                    8 x 12.5 = 100

9.) 20 ÷ 1.25   8 x 20   = 160   = 16 Ans.
                    8 x 1.25 = 10

20. Division by 9.9, 19.9, 39.9, 89.9, etc.

Rule: Raise the divisor by one-tenth (.1) to the higher divisor with
one significant digit. Divide the number by the divisor. Divide the quotient by the divisor again and add to answer as
a decimal.
For the lower numbers such as 9.9 and 19.9, it may be necessary to divide the decimal by the divisor again for accuracy
to more decimal places.

          1.) 7200 ÷ 29.9
               7200 ÷ 30 =   240
               240 ÷ 30 = +   .80
                           240.80

                              .26
                           240.8026 Ans.

The correct answer to two decimal places is 240.80. Divide the .8
by 30 and add as decimal. Accurate to four places.

          2.) 84200 ÷ 79.9
               84200 ÷ 80 =   1052.5
               1052.5 ÷ 80 =   1.315
                              1053.815 Ans.
               Accurate to two decimal places.

When the partial quotient 1052.5 has four digits for the whole
number, when divided by divisor, a decimal must be placed after
the first digit (left to right).

```
3.)  3792 ÷ 9.9
         3792 ÷ 10  =  379.2
         379.2 ÷ 10  =    3.792
         3.792 ÷ 10  =+   .03792
                      383.0299   or 383.03  Ans.
```

The correct answer is 383.030303

21. Division by 1½, 2½, 3½, etc.

Rule: Multiply the divisor by 2, 3, 4, or more to obtain a number with a single significant digit. This makes it easier to divide.

```
1.)  42 ÷ 1 1/2 or (1.5)
          double  84 ÷ 3  =  28 Ans.


2.)  92.5 ÷ 2.5
          double  185 ÷ 5  =  37  Ans.
```

Division by Other Fractions

```
3.)
  143 ÷ 3 1/3    3 1/3 = 10/3  =  10/3 x 3 =  10
            multiply by 3           429 ÷ 10 = 42.9   Ans.

4.)  754 ÷ 1 2/5  ( 1.4)     7/5 x 5 = 7
              multiply by 5   3770 ÷ 7 = 538.57   Ans.

5.)  856 ÷ 2 3/4      2 3/4 = 11/4     11/4 x 4 = 11

              856 x 4 = 3424
              3424 ÷ 11 = 311.272   Ans.
```

When dividing by fractions or mixed numbers the standard way, the dividend is multiplied by the fraction inverted.

```
6.)  750 ÷ 2 1/4       2 1/4 = 9/4

     750 x 4/9  =   3000/9 = 333.33  Ans.
```

22. By Trachtenberg Method (Simple)

Rule: Multiply the divisor by two as many times as required. Set down the sums and label each multiplication. Then subtract the highest sum from the dividend without exceeding dividend. Bring down the next digit and repeat. This is another form of common division.

129

1.)   3748 ÷ 62
  a.) 62   3748
  b.)+ 62 (2)- 372   6
    124   28   0.
  c.) +62 (3)   280
    186   - 248   4
  d.)+ 62 (4)   320
    248   - 310   5
  e.)+ 62 (5)   100
    310   - 62   1
  f.)+ 62 (6)
    372
           60.451 Ans.

2.)   3756592 ÷ 332
  a.)   332   3756592
      332   332   (1)
  b.)   664   436
      332   332   (1)
  c.)   996   1085
      332   996   (3)
  d.)   1328   499
      332   332   (1)
  e.)   1660   1672
          1660   (5)
          120
          .0   (0)
          1200
          996   (3)
          204
    Ans.   11315.03

Trachtenberg Fast Method of Division

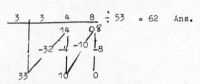

Step by Step explanation and procedure:
1. 5 goes into 33 six times. Set down 6 (in answer)
2. Bring down the 33.
3. Multiply 54 by 6 as follows:
   $6 \times 5 = 30, 6 \times 4 = 24; 30\ 24$
4. Add the 30 and the tens 2 = 32.
5. Subtract from 33 and set the 1 down.
6. Bring down the 4.
7. Subtract the units 4 of 2 4 from 14 and bring down the 10.
8. 5 goes into 10 twice. Set down the 2.
9. Multiply 54 by 2. 2×5 and 2×4 or 1 0 0 8
10. Subtract the 10 + 0 from 10 and set down 0.
11. Bring down the 8.
12. Subtract the units 0 8 from this. Remainder is 0.
    Division is complete.

If there was a remainder, the division would be carried out into decimal places.

$$3352 \div 54 = 62.074$$

Carrying out the division there would be a remainder of 4. proceed as follows. (All answers will be to the right of the decimal point.)

4   0   0   0   0   − 54 = .074 Ans.

1. 54 will not go into 40. Set down a 0. Bring down the 40.
1. 54 will go 7 times into 400. Multiply 54 by 7 as follows.
   $7 \times 5 = 35$; $7 \times 4 = 28$; 3 5 2 8. Add the 35 to tens 2; = 37
3. Subtract from 40. Set down the 3 and bring down a zero.
4. Subtract the units 8 from 30 and set down the 22.
5. 5 will go into 22 four times. Set down the 4. Multiply 54 by 4 for 2 0 1 6.
6. Subtract the 20 + 1 from 22 and set down the 1. Bring down a zero. Subtract the 6 from the 10 and set down the 4.
7. 5 will not go into 4, etc., etc.

Divisibility Rules

A number is divisible:
a. By 2, 4, 8 if its last digit, last two digits, last three digits are divisible by 2, 4, 8.

    1.) 54   2.) 34516   3.) 1356728
       ÷ 2   ÷ 4   ÷ 8

b. By 3 or 9 if the cross sum is divisible by them.

    1.) 456   cross sum = 4 + 5 + 6 = 15: 15 is divisible by 3.

    2.) 48735   cross sum = 4+8 + 7 + 3 + 5 = 27: divisible by 9 or 3.

c. By 5, if its last digit is 0 or 5.
d. By 25, if the last two digits are 25, 50, 75, or 00.
e. By 15, if it is divisible by 3 and 5.

    1.) 845670 is divisible by 3 or 5.

f. By 24, if it is divisible by 3 and 8.

    1.) 86088   2.) 86088   3.)   86088
      ÷ 3   ÷ 8   ÷ 24

131

g. By 11, if the eleven remainder (ER) is 0.
   (Subtract the digits in the even places of the number by the digits in the odd places to obtain the ER).

   1.) 3 9 2 3 7   odd places marked by dots.
        •   •   •
                         (9 + 3)- (3 + 2 • 7)
                          12 -   12 = 0

h. By 7. Subtract all digits to left from the three digits on right. If
   the difference is divisible by 7, the number is divisible by 7.

   1.) 72499   499   2.) 241969   969
               - 72   - 241
                 427   728

        divisible by 7   divisible by 7
                                or

Double the unit digit and   2 5 0 4|6
subtract it from the rest of   1 2   × 2
the number. Repeat until you   - 2 4 9|2
can see if the result is divisi-   -   4   × 2
ble by 7.   2 4|5   × 2
                                   -      1 0
                                      1 4 is ÷ by 7

                                    1 6 2 3 6
                                        236
i. By 11. Subtract all digits   -   16
   to left from the   220
   three digits on   which is divisible
   right. If the differ-   by 11.
   ence is divisible by
   11, the number is   09354   13 ?
   divisible by 11.   9 8 5|4
                                  +   1 6   × 4
j. By 13. Multiply the units   1 0 0|1
   digit by 4 and add   +   4   × 4
   to the rest of the   1 0 4 is ÷ 13
   number. Repeat.
                                   0573 - 17
k. By 17. Multiply units   9 6 7|3
   digit by 5 and   -   1 5   × 5
   subtract from   9 5|2
   rest of the num-   -   1 0   × 5
   ber. Repeat.   8 5 is ÷ by 17

l. By 19. Double the units   10013 - 19
   digit and add it to   1 0 0|3
   the rest of the   +   6   × 2
   number. Repeat.   1 0 0|7
                                +   14
                                  1 1 4
                                +   8
                                  1 9 ÷ 19 = 1

Rule: Multiply the divisor by a number which is as close to 100 as possible, either over or under. Then use the difference to multiply the dividend unit and either add or subtract from the rest of the answer.

$\div$ 7 $\qquad \dfrac{7 \times 14 \mid -2}{100}$ $\quad$ 7 x 14 = 98 (2 under). Subtract

$\div$ 11 $\qquad \dfrac{11 \times 9 \mid -1}{100}$ $\quad$ 11 x 9 = 99 (1 under). Subtract

$\div$ 13 $\qquad \dfrac{13 \times 8 \mid + 4}{100}$ $\quad$ 13 x 8 = 104 (4 over). Add

$\div$ 17 $\qquad \dfrac{17 \times 6 \mid -2}{100}$ $\quad$ 17 x 6 = 102 (2 over). Add

$\div$ 15 and $\div$ 19 are not applicable

Checking Division

Standard Method By Multiplication.

$$276 \div 8 = 34.5$$
$$34.5 \times 8 = 276$$

9-Test

The CS and the 9-Test were explained under multiplication and checking multiplication.

Rule: Multiply the CS of the quotient by the CS of the divisor. Add the CS of the remainder. The resulting CS should agree with the CS of the dividend. Or: Quotient $\times$ divisor + remainder = dividend

2432 $\div$ 76 = 32

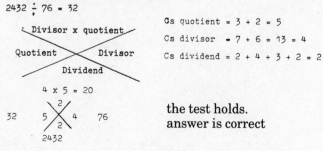

Cs quotient = 3 + 2 = 5

Cs divisor = 7 + 6 = 13 = 4

Cs dividend = 2 + 4 + 3 + 2 = 2

the test holds.
answer is correct

133

As in multiplication, this test does not reveal errors of position and errors of 9, but it is simple and is of great value in spotting mistakes.

POWERS OF TEN

Introduction

Where the accuracy of calculation is not critical, rounding off and using powers of 10 can greatly facilitate calculations. This method is extremely useful in many phases of engineering and science, especially in electronics, where billions are to be multiplied by millions.

Below is the table for powers of ten for whole numbers and for decimals. The table starts with 10 raised to the zero power of 10 = 1. Any number raised to the zero power is equal to 1.

$$
\begin{aligned}
1 &= 10^0 \\
10 &= 10^1 \quad (10 \times 1) \\
100 &= 10^2 \quad (10 \times 10) \\
1000 &= 10^3 \quad (10 \times 10 \times 10) \\
10000 &= 10^4 \quad (10 \times 10 \times 10 \times 10) \\
100000 &= 10^5 \quad (10 \times 10 \times 10 \times 10 \times 10) \\
1000000 &= 10^6 \quad (10 \times 10 \times 10 \times 10 \times 10 \times 10) \\
\text{etc,} & \quad \text{etc,} \\
1/10 &= 10^{-1} = .1 \\
1/100 &= 10^{-2} = .01 \\
1/1000 &= 10^{-3} = .001 \\
1/10000 &= 10^{-4} = .0001 \\
1/100000 &= 10^{-5} = .00001 \\
1/1000000 &= 10^{-6} = .000001 \\
\text{etc,} & \quad \text{etc,}
\end{aligned}
$$

Converting To Powers of Ten

Rule: For whole numbers, starting from the right hand side, (the decimal point is not shown but it is assumed to be there) count off one place for each time you move the decimal point to the left. You can move the decimal point

137

as many places as you desire but the standard method is to stop with a single whole integer to the left of the decimal.

$$2.5000 = 2.5 \times 10^4$$

(25000)

The decimal point was moved four places to left. Note as 10^4 You may stop anywhere but the above is the most practical. Other examples:

$$2\ 5.000 = 25 \times 10^3$$
$$2\ 5\ 0.00 = 250 \times 10^2$$
$$.25000 = .25000 \times 10^5$$
$$.025000 = .02500 \times 10^6$$

Note that in the last example a zero must be used to indicate the empty place. The same applies to decimals but you move the decimal point to the right and indicate the number of places with a negative sign.

$$.00003 = 3 \times 10^{-5}$$
$$.0000075 = 7.5 \times 10^{-6}$$

Do not write as 7.5^{-6}. Use the 10^{-6} notation.

Rounding Off Numbers

Many times it is not necessary to use a number such as 782,948. We can round off this number to two or three places. To round off to two places, write 780,000, the 2948 is replaced by four zeros. To round off to three places write 783,000. Since the integer following the 2 is larger than 5, raise the 2 to the next higher digit 3.

Rule: If the integer to be dropped is less than 5, just change it to a 0. If it is more than 5, increase the integer by 1. If it is 5, increase the adjacent integer by one or not. It will average out. Decimals are rounded off in the same manner.

$$.034287 = .034 \text{ to two places}$$
$$= .0343 \text{ to three places}$$

(change 2 to 3 since the following integer is 8.)

Examples:

		two places		three places
1.)	333479	= 330,000	or	333,000
2.)	475300	= 470,000	or	475,000
3.)	6784742	= 6,800,000	or	6,780,000
4.)	.00007643	= .000076	or	.000764
5.)	.004956	= .0049	or	.00495

The .0049 can be written as .005

As Power Of Ten

	two places		three places
1.)	3.3×10^5	or	3.33×10^5
2.)	4.7×10^3	or	4.75×10^5
3.)	6.8×10^6	or	6.78×10^6
4.)	7.6×10^{-5}	or	7.64×10^{-5}
5.)	4.9×10^{-3}	or	4.96×10^{-3}
	or 5×10^{-3}		

Multiplication with Powers of 10

Rule: Add the powers. Multiply the integers.

$(3 \times 10^3)(2 \times 10^5) = (3 \times 2) \times 10^{3+5} = 6 \times 10^8$ Ans.

$(4.4 \times 10^6)(3 \times 10^6) = (4.4 \times 3) \times 10^{6+6} = 13.2 \times 10^{12}$ Ans.

Negative Powers

$(3 \times 10^{-3})(2 \times 10^{-4}) = (3 \times 2) \times 10^{-3-4} = 6 \times 10^{-7}$ Ans.

$(4.7 \times 10^{-5})(3 \times 10^{-6}) = (4.7 \times 3) \times 10^{-5-6}) = 14.\underline{1} \times 10^{-11}$ Ans.

and $14.1 \times 10^{-11} = 1.4 \times 10^{-12}$ Ans.

Mixed Powers

$(4 \times 10^6)(2 \times 10^{-2}) = 8 \times 10^{6-2} = 8 \times 10^4$ Ans.

$6 + (-6) = 6 - 3 = 3$

$(3 \times 10^3)(3 \times 10^{-6}) = 9 \times 10^{-3}$ Ans.

$3 + (-6) = 3 - 6 = -3$

$(4 \times 10^{-6})(2 \times 10^6) = 8 \times 10^{-6+6} = 8 \times 10^0 = 8$

Examples:

32,400 x 989.246

$(3.2 \times 10^4) \times (9.9 \times 10^5)$ = 31.68×10^9

or 3.17×10^{10}

The above calculation is accurate to .1%. Even pocket calculators round off large numbers and answers.

$$475,582 \times 3,349$$
$$(4.8 \times 10^5) \times (3.3 \times 10^3) = 5.94 \times 10^9 \quad \text{Ans.}$$
$$= 1,594,000,000 \quad \text{Ans.}$$

Imagine doing this the long way! The actual answer is 15,930,590,018. the above answer is approximately .6% off (below).

$$.000029 \times 753,000$$
$$(3 \times 10^{-5}) \times (7.5 \times 10^5)$$
$$= 22.5 \times 10^{-5+5} = 22.5 \times 10^0 = 22.5 \quad \text{Ans.}$$

Note that the decimal was converted to 3.

Finally:
$$(3.3 \times 10^{-18}) (4 \times 10^{24})$$

In English:3.3 million, million, millionths times 4 million, billion, billion.

Division with Powers of 10

Rule: The whole integers are divided normally while the powers are subtracted. (Remember that a negative number is subtracted by changing its sign and adding the number.)

1.) $(8 \times 10^4) \div (2 \times 10^2) = 8 \div 2$ and $10^{4-2} = 4 \times 10^2$ Ans.

2.) $\dfrac{6 \times 10^6}{2 \times 10^2} = (6 \div 2)$ and $10^{6-2}) = 3 \times 10^4$ Ans.

3.) $\dfrac{7 \times 10^{-3}}{2 \times 10^{-3}} = (7 \div 2)$ and $10^{-3-(-3)} = 3.5 \times 10^{-3+3}$
$$= 3.5 \times 10^0 = 3.5 \quad \text{Ans.}$$

4.) $\dfrac{16 \times 10^{-4}}{2 \times 10^{-5}} = 8 \times 10^{-4-(-5)} = 8 \times 10^{-4+6} = 8 \times 10^2$ Ans.

5.) $\dfrac{6.5 \times 10^{-6}}{2 \times 10^4} = 3.25 \times 10^{-6-4} = 3.25 \times 10^{-10}$ Ans.

6.) $\dfrac{4 \times 10^2}{2 \times 10^6} = 2 \times 10^{2-6} = 2 \times 10^4$ Ans.

7.) $\dfrac{800,000}{2000} = \dfrac{8 \times 10^5}{2 \times 10^3} = 4 \times 10^{5-3} = 4 \times 10^2$ Ans.

8.) $\dfrac{334,200}{.0002} = \dfrac{3.3 \times 10^5}{2 \times 10^{-4}} = 1.65 \times 10^{5-(-4)} = 1.65 \times 10^9$ Ans.

Squaring with Powers of 10

Rule: Square the integer. Double the power.

1.) $(4 \times 10^3)^2$ = (4×4) and $(10^{3 \times 2})$ = 16×10^6 or 1.6×10^7
Ans.

2.) $(75 \times 10^5)^2$ = $75^2 \times 10^{12}$ 5625 $\times 10^{12}$ o4 5.63 $\times 10^{15}$ Ans.

3.) $(8 \times 10^{-2})^2$ = $8^2 \times 10^{-2 \times 2}$ = 64×10^{-4}

Notes: In America a trillion is 1×10^{12}.
In England a trillion is 1×10^{18}.
A billion in America is 1×10^9.
In England it is a million million or 1×10^{12}.

Bibliography

Berkeley, Edmund C., A Guide To Mathematics For The Intelligent Nonmathematician, Simon and Schuster, N.Y.

Cutler, Ann, and McShane, Rudolph, The Trachtenberg Speed System Of Basic Mathematics, Doubleday & Co. Inc., Garden City, N.Y. 1960.

Mennenger, Karl, Calculator's Cunning, Basic Books, Inc. N.Y., 1961.

Meyers, Lester, Pocket Books Incorporated, N.Y. High Speed Math, 1947.

Simon, William, Mathematical Magic, Charles Scribner's Sons, New York, 1964.

Stisker, Henry, How To Calculate Quickly, Dover Publications, Inc. 1945.